Computer Aided Methods in Optimal Design and Operations

Series on Computers and Operations Research

Series Editor: P. M. Pardalos *(University of Florida)*

Published

Vol. 1 Optimization and Optimal Control
eds. P. M. Pardalos, I. Tseveendorj and R. Enkhbat

Vol. 2 Supply Chain and Finance
eds. P. M. Pardalos, A. Migdalas and G. Baourakis

Vol. 3 Marketing Trends for Organic Food in the 21st Century
ed. G. Baourakis

Vol. 4 Theory and Algorithms for Cooperative Systems
eds. D. Grundel, R. Murphey and P. M. Pardalos

Vol. 5 Application of Quantitative Techniques for the Prediction of Bank Acquisition Targets
by F. Pasiouras, S. K. Tanna and C. Zopounidis

Vol. 7 Computer Aided Methods in Optimal Design and Operations
eds. I. D. L. Bogle and J. Žilinskas

Forthcoming

Vol. 6 Recent Advances in Data Mining of Enterprise Data: Algorithms & Applications
eds. T. Warren Liao and Evangelos Triantaphyllou

Series on Computers and Operations Research Vol. 7

Computer Aided Methods in Optimal Design and Operations

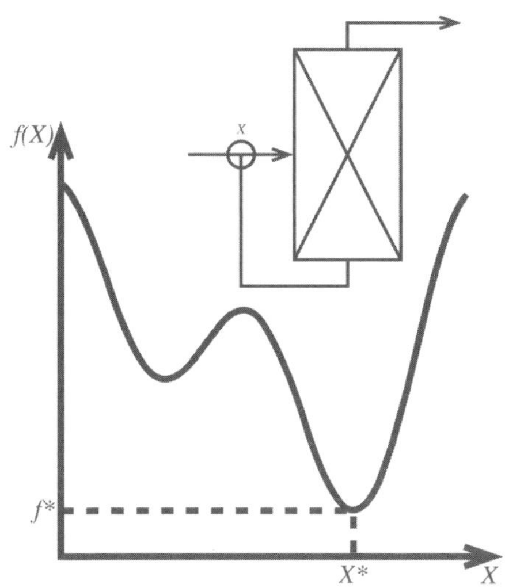

Editors

I D L Bogle
University College London, UK

J Žilinskas
Institute of Mathematics and Informatics, Lithuania

World Scientific

NEW JERSEY • LONDON • SINGAPORE • BEIJING • SHANGHAI • HONG KONG • TAIPEI • CHENNAI

Published by

World Scientific Publishing Co. Pte. Ltd.

5 Toh Tuck Link, Singapore 596224

USA office: 27 Warren Street, Suite 401-402, Hackensack, NJ 07601

UK office: 57 Shelton Street, Covent Garden, London WC2H 9HE

British Library Cataloguing-in-Publication Data
A catalogue record for this book is available from the British Library.

COMPUTER AIDED METHODS IN OPTIMAL DESIGN AND OPERATIONS
Series on Computers and Operations Research — Vol. 7

ISBN 981-256-909-X

Printed in Singapore by World Scientific Printers (S) Pte Ltd

Preface

This book contains papers presented at the bilateral workshop of British and Lithuanian scientists “Optimal process design” held in Vilnius, Lithuania from 15th to 17th of February, 2006. The workshop was supported by the British Council through the INYS programme.

The workshop was organized by UCL (University College London), UK, and the Institute of Mathematics and Informatics, Lithuania. The meeting was co-ordinated by Professor A. Žilinskas and Dr J. Žilinskas from the Institute of Mathematics and Informatics, and Professors E. S. Fraga and I. D. L. Bogle from UCL.

The British Council International Networking for Young Scientists Programme (INYS) brings together young researchers from the UK and other countries to make new contacts and promote the creative exchange ideas through short conferences. Mobility for young researchers facilitates the extended laboratory in which all researchers now operate: it is a powerful source of new ideas and a strong force for creativity. Through the INYS programme the British Council helps to develop high quality collaborations in science and technology between the UK and other countries and shows the UK as a leading partner for achievement in world science, now and in the future. The INYS programme is unique in that it brings together scientists in any priority research area and helps develop working relationships. It aims to encourage young researchers to be mobile and expand their knowledge. The homepage of the INYS supported workshop “Optimal Process Design” is available at http://www.mii.lt/inys/.

The workshop was divided into four sections: General Methodologies in Design, Design Applications, Visualization Methods in Design, and Operations Applications. Twenty two talks were selected from twenty seven submissions from young UK and Lithuanian researchers. Professor

C. A. Floudas from Princeton University, USA, gave an invited lecture. Some review lectures were also given by the other members of the scientific committee.

This book contains review papers and revised contributed papers presented at the workshop. All papers were reviewed by leading scientists in the field. We are very grateful to the reviewers for their recommendations and comments.

We would like to thank the British Council for financial and organizational support.

We hope that this book will serve as valuable reference document for the scientific community and will contribute to the future co-operation between the participants of the workshop.

I. D. L. Bogle
J. Žilinskas

Contents

HYBRID METHODS FOR OPTIMISATION

E. S. FRAGA

Centre for Process Systems Engineering, Department of Chemical Engineering University College London (UCL), London WC1E 7JE, United Kingdom

Computer aided design tools for industrial engineering typically require the use of optimisation. The optimisation problems in industrial engineering are often difficult due to the use of nonlinear and nonconvex models combined with underlying combinatorial features. The result is that no single optimisation procedure is typically suitable for most design tasks. Hybrid procedures are able to make use of the best features of any method while ameliorating the impact of the disadvantages of each method involved. This paper presents an overview of hybrid methods in engineering design. A simple case study is used to illustrate one hybrid optimisation procedure.

1. Introduction

Computers are used in industrial engineering throughout the whole life cycle. At the early stages of the cycle, computer aided design tools are used to identify good or promising design alternatives. Subsequently, further tools are used to refine these alternatives using more complex models as information becomes available and issues must be addressed. The earlier issues can be addressed, the greater the likelihood that the final design generated meets the criteria imposed on it (economic, environmental, societal). Therefore, there is constant pressure to have as complex a model as possible for the design problem under consideration as early as possible.

This constant pressure is resisted by the need for more powerful and capable optimisation tools to handle the increased complexity. Optimisation forms the core of many computer aided engineering tools. The types of optimisation models used in industrial engineering range from linear programming through to mixed integer differential/integral nonlinear programming. Generic technologies have been developed for most classes of optimisation problems with varying success. Commercial software is available, including, for instance, the set of solvers available through the NEOS server.[1]

2. Hybrid Methods for Optimisation

Although there has been significant progress in the development of generic solvers, many problems in industrial engineering cannot be handled with these solvers. Models in industrial engineering, especially those in the processing industries, often exhibit nonlinear, nonconvex and discontinuous behaviour. Furthermore, the models may pose inherent numerical difficulties for computational tools due to behaviour in the limits of the domains of the variables (e.g. the *log*-mean temperature difference equation in heat exchanger design) or are valid only in a restricted domain and may be meaningless outside that domain (e.g. mole fractions). In some cases, models may also exhibit noise (e.g. due to online experimental measurements as part of the models used).

Therefore, for many industrial engineering applications, targeted optimisation procedures are developed. These targeted procedures are often based on stochastic methods, including evolutionary programming methods, such as genetic algorithms,[2] and simulated annealing.[3] The appeal of this class of methods is their ease of implementation and their robustness with respect to the issues mentioned above, making them suitable for use by non-experts in the area of optimisation. Their greatest disadvantage, however, is the number of parameters that require setting and for which values are often difficult to ascertain based purely on the problem considered.

Although these stochastic methods can be successful in identifying good solutions, they often do not achieve the best solutions possible and also do not necessarily provide any insight into how far from the best the solutions obtained may be. The advantage of the more traditional, mathematical programming, approaches is that they can address some of these issues. Therefore, one reasonable approach is to consider the development of *hybrid* procedures that combine the best attributes of these classes of methods.

Hybrid methods are so called because they combine one or more methods to work together in solving a given problem:

> Hybrid (*Hy"brid*), a.
> derived by a mixture of characteristics from two distinctly different sources;[4]

There are two ways to combine two or more methods: sequential or embedded. Examples of both are presented in what follows.

3. Embedded Hybrid Methods

In an embedded method, an outer procedure is used to determine the values of all the decision variables or, possibly, a subset of these variables. An inner procedure is then invoked with the values determined by the outer procedure either to determine the values of the remaining decision variables or to refine the values of all the decision variables determined by the outer procedure. Once the inner procedure has completed, the outer procedure is given control again and another iteration performed until the appropriate stopping criterion is met.

The simplest example of an embedded hybrid method appears in the various modifications to the conjugate gradient method for handling the line search. The outer method determines a search direction and the inner procedure manipulates the decision variables subject to remaining on a line defined by the search direction. However, this example is arguably not a hybrid procedure in that the outer procedure does not actual define any values for the decision variables.

In computer science applications, a large number of what are known as *neighbourhood search* or *local search* algorithms have been developed. These are typically a combination of a backtracking algorithm used to search through a graph based representation of the solution space with a local embedded search procedure used to determine the best alternative path to choose at any point in the forward traversal. See Ahuja *et al.*[5] for a general survey of these types of methods.

Shouraki & Haffari[6] describe their experiences with different local search algorithms within the STAGE procedure for tackling combinatorial problems. STAGE combines local search methods with backtracking procedures, preserving the scalability of the local search methods while aiming for the exhaustive properties of backtracking methods. Prestwich[7] describes the Incomplete Dynamic Backtracking method which combines local search with backtracking so as to preserve the advantages of both approaches without losing the scalability of the local search methods. More recently, van Hentenryck & Michel[8] present a formulation for describing hybrid search procedures based on backtracking and local or neighbourhood search methods.

A more general embedded hybrid optimisation approach is the incorporation of local search or refinement techniques within a stochastic global optimisation procedure. Locatelli & Schoen[9] describe formally how a local search algorithm affects a global optimisation procedure. They apply such a method to the minimisation of potential energy as modelled by the Lennard-Jones equation.

Frequently, the stochastic optimisation procedure is a genetic algorithm[2] (GA) or simulated annealing[3] (SA). Some examples of such approaches are described in the remainder of this section.

Thomsen[10] describes the effects of incorporating a local search within a genetic algorithm to define a *Lamarckian* GA. A Lamarckian GA is one in which population members can be modified by a local search procedure (cf. the distinction between Lamarckian and Darwinian evolution[11]). Three approaches are compared: one without any local search, one where the current best solution is refined and one where a randomly chosen solution from the current population is chosen.

Ganesh & Punniyamoorthy[12] describe a combined GA and SA procedure where, at the end of each generation of a genetic algorithm, each member of the current population is used as an initial guess for a simulated annealing procedure. The results of all the SA applications are used to define a new population (using standard selection procedures in the GA). A similar approach was used by Ponnambalam & Reddy[13] for integrating lot sizing and sequencing in flow-line scheduling.

The local search procedure need not be deterministic. For instance, Tulpan & Hoos[14] present a stochastic local search method based on a random walk procedure which has been extended with a local search procedure which resolves conflicts (i.e. constraint violations). They have applied this method to the DNA code design problem.

Theos *et al.*[15] describe the PANMIN program which is based on two stochastic global optimisation methods which use local searches as intermediate steps and for refinement of solutions. One of the methods is similar to a controlled random search.[16] The second method implements a topographical multilevel single linkage approach. This has some similarity to a controlled random search but with memory. The method also uses a Bayesian[17] statistical method to provide a stopping criteria by estimating the number of global minimisers in the domain. The local search, which forms part of the core algorithm, uses the Merlin system[18] which provides links to a number of local search methods include both direct search and gradient based methods.

Alternative methods for the outer global optimisation procedures have also been developed. Smyth *et al.*[19] combine a tabu search with iterated local search. Jussien & Lhomme[20] also combine a tabu search with a local search procedure, this time a search over partial assignments, instead of complete assignments, for open-shop scheduling problems.

Recently, another stochastic approach derived from observation of biology, based on analogies with ant colonies or particle swarms, has been investigated. For hybrid methods, Meyer & Ernst[21] combine an *ant colony optimisation* (ACO) model with constraint propagation to tackle problems with hard constraints that would otherwise be inappropriate for ant colony models. Lee & Lee[22] combine GA, ACO and heuristics to solve resource allocation models.

The commonality of all these methods is the combination of an outer global optimisation procedure with a targeted local search method. The aim is to enhance the convergence of the outer procedure using the fine tuning capabilities of local search methods. Without this tuning, many of the global optimisation methods used may converge to the global optimum in theory but in practice achieve less spectacular results.

Before continuing on to the other form of hybrid optimisation, it is worth noting that not all embedded approaches embed a local search method within a global optimisation procedure. In fact, Fraga & Žilinskas[23] present a family of embedded hybrid methods for the optimal design of heat-integrated process flowsheets in which the outer method is a direct search local optimisation procedure and the embedded method is a genetic algorithm. This particular combination is chosen because of the decomposition used for the process model. The outer procedure handles the NLP aspects whereas the inner procedure takes care of the combinatorial elements. The particular combination is shown to be highly effective and efficient.

4. Sequential Hybrid Methods

The procedures presented in the previous section demonstrate the wide range of applicability of the embedded form of hybrid optimisation. However, the alternative approach for combining optimisation procedures is more straightforward and can still achieve significant improvements over the use of a single method. In a sequential approach, one method is applied and a solution, or possibly a set of solutions, is generated. This solution or set of solutions is then used as the initial guess for a subsequent method. The solution from the second step can be fed into yet another method or back into the first method, forming the basis of an iterative procedure. These sequential hybrid methods are also known as *multi-start* algorithms[24] although, for some authors, multi-start methods imply a single method with multiple attempts using different initial guesses.

In principle, any combination of methods can be used. For instance, very recent work by Xia & Wu[25] presents a sequential hybrid procedure using a particle swarm optimisation (PSO) method to initialise a simulated annealing procedure. Fraga & Papageorgiou[26] use an interval analysis based stochastic procedure to provide feasible or close to feasible initial solutions for the following mathematical programming stage for the design and optimisation of water distribution networks.

Instead of attempting to enumerate further even a small number of such approaches, the rest of this paper is devoted to a simple case study which demonstrates the potential benefits of using sequential hybrid methods.

5. Illustrative Case Study

A process plant will typically have large cooling and heating demands. For instance, a popular technology for separating liquid mixtures is distillation. A distillation unit operates by boiling liquid at the bottom of the unit and condensing vapour at the top. Meeting the heating and cooling requirements can involve large amounts of utilities, such as steam and cooling water. Besides the obvious economic impact, there are also significant environmental issues from utility consumption. Therefore, it is beneficial to reduce utility consumption whenever possible.

Utility consumption can be reduced by using excess heat in one part of a process plant to meet the heating requirements elsewhere in the same process plant, subject to the laws of thermodynamics. Using heat in this way is known as process integration. Identifying the optimum integration between all the processing units in a process plant is known as the *heat exchanger network synthesis* (HENS) problem.

The definition of a HENS problem is a set of cold streams, a set of hot streams and the set of utilities available for meeting any cooling and heating demands not satisfied by integration. Mathematically, the aim is to minimise, for instance, an annualised cost for meeting the heating and cooling requirements of a process plant taking into account not only the utility consumption but also the cost of equipment.

As an optimisation problem, all possible integrations must be considered. This is a combinatorial problem and is particularly challenging when we allow for streams to be split so that, for instance, a hot stream may exchange heat with two cold streams in parallel. Previous attempts at solving the full heat exchanger network synthesis problem with stream splitting have been based on the *a priori* definition of a superstructure.[27,28]

For larger problems, an efficient superstructure can be difficult to generate. By efficient, in this case, we mean a superstructure that contains hopefully all solutions of interest with minimal coverage of solutions that are less likely to be good. A tighter superstructure will lead to easier to solve optimisation problems, in some cases making the difference between a problem which is solvable and one which is intractable. Recently, with this aim, we have developed a multiple ant colony model approach for identifying a suitable superstructure as the first step in a multi-step sequential hybrid optimisation method.[29] In what follows, we illustrate the hybrid procedure used to solve the nonlinear programme defined by the superstructure generated by this ant colony method.

Table 1. Heat exchanger network synthesis case study.

Process Streams				
Stream	T_{in} (°C)	T_{out} (°C)	Q (kW)	h $\left(\frac{kW}{K \cdot m^2}\right)$
H1	200	40	6400	0.8
H2	120	60	600	0.8
H3	90	50	200	0.8
C1	25	180	3100	1.6
C2	80	210	3250	1.6
C3	35	160	2250	1.6
Utilities				
Type	T_{in} (°C)	T_{in} (°C)	h $\left(\frac{kW}{K \cdot m^2}\right)$	c_u $\left(\frac{£}{kW \cdot y}\right)$
Steam	220	219	1.6	700
Water	30	40	0.8	60

Note: Q is the amount of heating or cooling required for each stream, h is the heat transfer coefficient for each process stream and each utility, and c_u is the cost of each utility.

The problem we consider is a generalisation of the stream splitting case study presented by Morton,[30] shown in Table 1. The resulting superstructure identified by the ACO step,[29] and which forms the basis of the subsequent optimisation steps, is shown in Fig. 1.

The nonlinear programming model has 13 continuous variables: 7 heat exchanger duties and 6 split fractions. Heat exchanger duties are represented by `Xmn` in Fig. 1, where `m` indicates the cold stream index and `n`

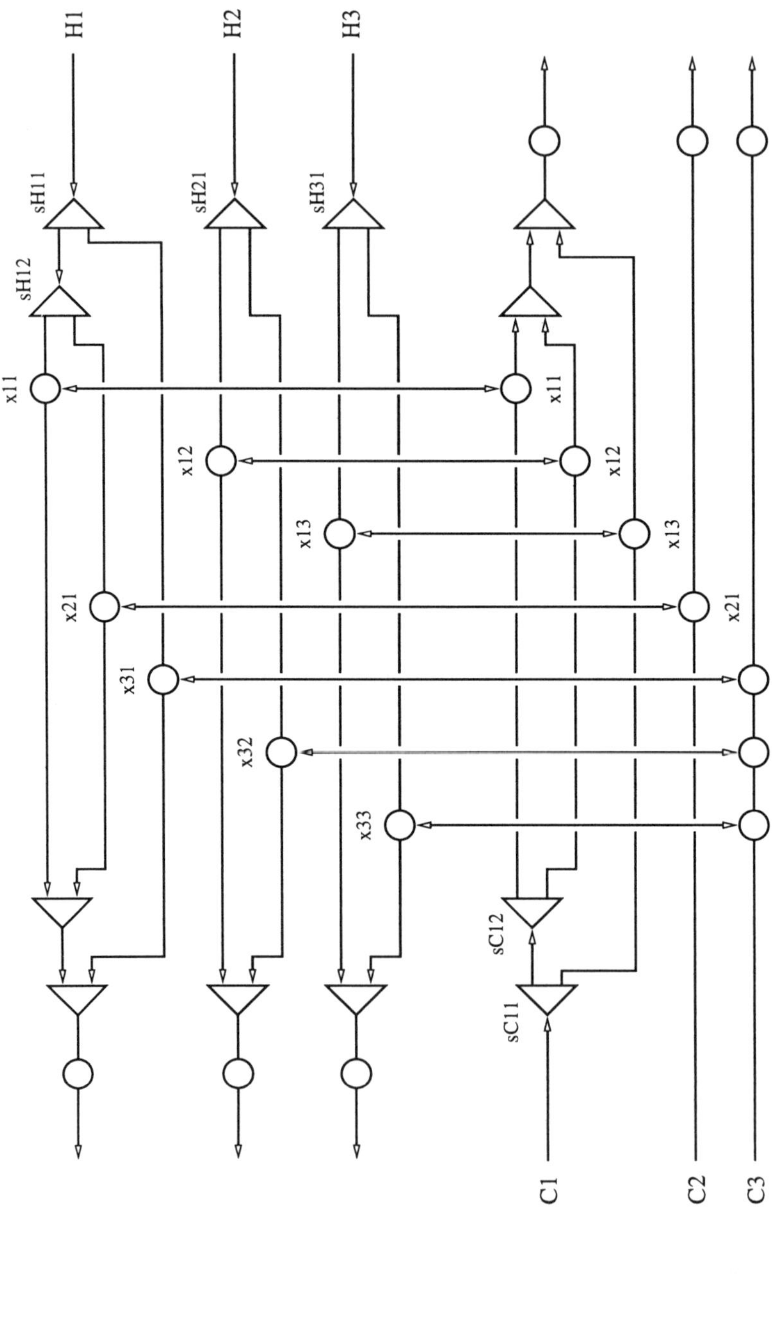

Figure 1. Superstructure for Morton case study obtained using an ant colony optimisation procedure.

the hot stream index. The split fractions are represented by `sHab` for hot streams and `sCab` for cold streams, where `a` is the index of the hot or cold stream and `b` is a counter to ensure unique labels for these splitters. All exchange variables are normalised so that all the variables take values $\in [0, 1]$.

The exchange variables represent the amount of exchange as a fraction of the maximum possible for that particular match. For a given match, the maximum possible is the minimum of the amounts available on each stream involved. The amounts available depend on the values of the split fractions. For instance, the match between cold stream `C2` and hot stream `H1`, indicated by `x21` in the superstructure, would have a maximum amount $Q_{\max}$ defined as:

$$Q_{\max} = \min \{Q_{C2}, \mathtt{sH11} \times (1 - \mathtt{sH12}) \times Q_{H1}\} \tag{1}$$

where $Q_{C2} = 3250$ kW and $Q_{H1} = 600$ kW, as given in Table 1, and where it is assumed that the split fraction variables specify what proportion of the duty goes to the upper stream as shown in the superstructure diagram.

This formulation is used to provide a feasible search space which is large in comparison with the full domain $\underline{x} \in [0, 1]^{13}$. Having the feasible space as large as possible with respect to the variable bounds makes searching more effective for stochastic methods. A second advantage of this formulation is that the point $\underline{x} = \underline{0}$ is a feasible point, one which corresponds to the non-integrated solution and which is a good starting point for any search from an engineering perspective.

The procedure for the solution of this NLP is shown in Algorithm 1. This procedure iteratively applies a number of direct search methods, specifically the Hooke & Jeeves, Implicit Filtering and BFGS methods from Kelley,[31] followed by two stochastic methods, a genetic algorithm and a simulated annealing procedure. The direct search methods are applied in parallel and the best of the results used as the initial population, size 1, for the genetic algorithm. The best solution obtained by the GA is used as an initial guess to a simulated annealing procedure. Finally, the best solution obtained by the SA is presented to the engineer for direct fine tuning, if desired. The result is then compared with the initial starting guess before the direct search methods were applied and convergence determined.

This iterative multi-start procedure uses direct search for fine tuning and the two stochastic methods for global searching. The results obtained for this case study are shown in Table 2 with the final heat exchanger network presented in Fig. 2 with the exchangers annotated with the actual

Algorithm 1 Multi-start hybrid procedure for solving the heat exchanger network synthesis problem.

Let $\underline{x} \leftarrow \underline{0}$ ▷ Guaranteed feasible point
converged $\leftarrow$ **false**
while not converged **do**
  $\underline{x}^{(1)} \leftarrow \underline{x}$
  improved $\leftarrow$ **true** ▷ For direct search only
  while improved **do**
    $\underline{x}^{(2)} \leftarrow$ best solution from direct search methods applied to $\underline{x}^{(1)}$
    improved $\leftarrow \left|\left|\underline{x}^{(2)} - \underline{x}^{(1)}\right|\right| > \epsilon$
    $\underline{x}^{(1)} \leftarrow \underline{x}^{(2)}$
  end while
  $\underline{x}^{(3)} \leftarrow$ GA : initial population $= \left\{\underline{x}^{(2)}\right\}$
  $\underline{x}^{(4)} \leftarrow$ SA : $\underline{x}^{(3)}$
  $\underline{x}^{(5)} \leftarrow$ engineer interaction!
  converged $\leftarrow \left|\left|\underline{x} - \underline{x}^{(5)}\right|\right| < \epsilon$
  $\underline{x} \leftarrow \underline{x}^{(5)}$ ▷ New iterate
end while

amounts exchanged, in kW.

The table shows how the values of each of the decision variables evolves during the iterative process. All the methods contribute to the improvement of the objective function at some point in the iteration and user interaction is used twice. User interaction is particularly useful for forcing variables to their bounds. The values changed by the user are indicated in bold in the table and without any trailing numbers after the decimal point. Any values before user interaction that appear to be at one of the bounds (e.g. "1.000") are in fact very close to the bound but not exactly on the bound.

Although it is often possible to design and implement optimisation procedures that attempt to push variables to bounds, this is difficult to do in a generic manner. The user, however, can often tell whether it is reasonable for a variable to actually be at the bound and, in this case, this proves to be true.

It is useful to analyse the evolution of a selection of the variables. Figure 3 shows graphically how these variables evolve. Specifically, we see that `x32` and `sH31` swing from one end of their domain to the other when the stochastic optimisation procedures are invoked, demonstrating the global search capabilities of these methods. We also see that variable `sH12` evolves

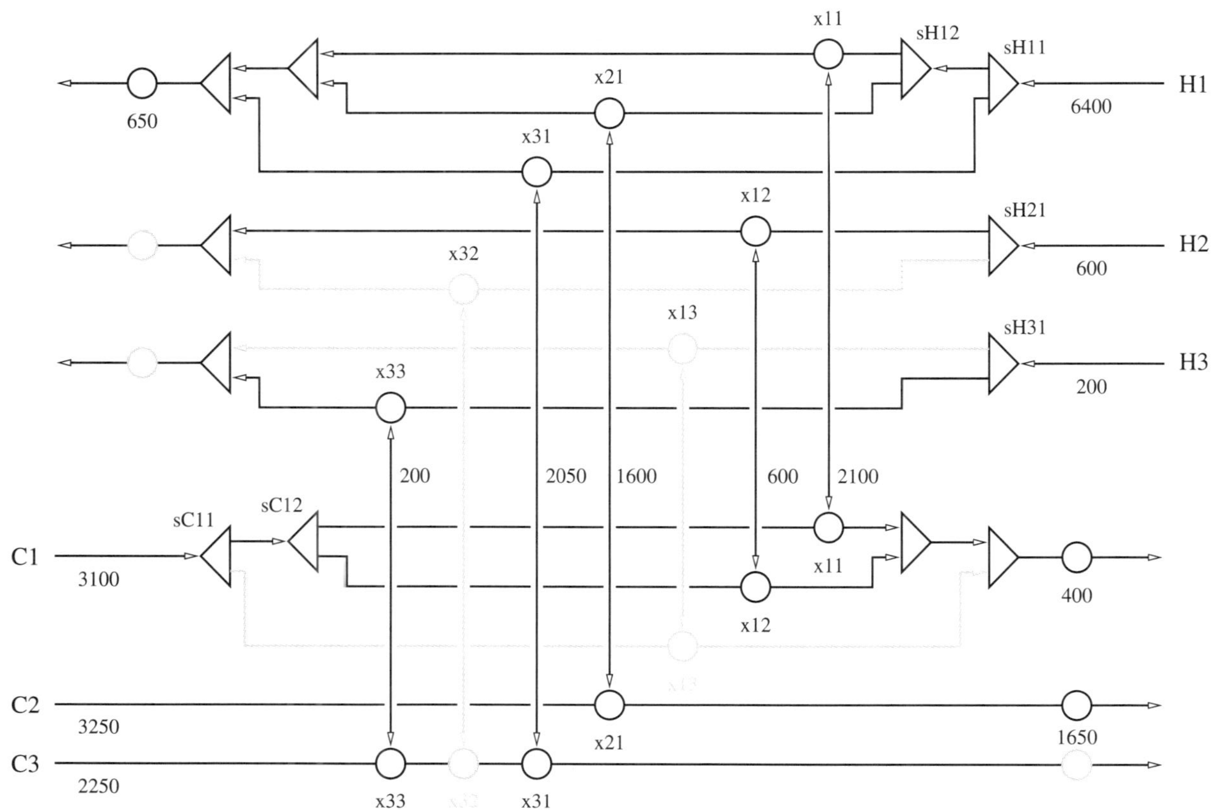

Figure 2. Final heat exchanger network design obtained for the case study. The elements of the original superstructure which are not present in the final network have been shown in light grey.

Table 2. Evolution of best solution during the application of the hybrid multi-start algorithm.

x_i	Step							
	Initial	IF	HJ	BFGS	GA+SA	UI	HJ	UI
x11	0.	.807	.998	1.000	1.000	1.	1.	1.
x12	0.	.697	.998	.999	1.000	1.	1.	1.
x13	0.	.695	.999	1.000	.998	1.	1.	1.
x21	0.	.806	.744	.744	.745	.745	.745	.745
x31	0.	.962	.998	.999	1.000	1.	1.	1.
x32	0.	.695	.999	.999	.320	.320	.320	.320
x33	0.	.694	.998	.998	1.000	1.	1.	1.
sH11	0.	.688	.649	.649	.663	.663	.663	.663
sH12	0.	.193	.373	.374	.489	.489	.493	.493
sH21	0.	.695	.999	1.000	1.000	1.	1.	1.
sH31	0.	.694	.999	.999	.039	.039	.004	0.
sC12	0.	.693	.599	.600	.674	.674	.678	.678
sC11	0.	.693	.834	.834	.994	1.	1.	1.
$f(\underline{x})$	3.306	1.211	.914	.912	.908	.893	.892	.888

slowly to its final value, indicating the contribution made by several methods in combination.

6. Discussion

This paper has presented a brief overview of hybrid methods for optimisation. The methods have been placed into two categories, embedded and sequential. A key distinction between the two is that the former tend to be defined and implemented for specific problems or problem areas whereas the latter can make use of generic implementations.

A case study on the design of a heat exchanger network has been presented to demonstrate the power of a hybrid optimisation procedure. In this case, a multi-start, or sequential, procedure has been used. This procedure combines both stochastic and deterministic methods and the case study shows the contribution made by both types of methods. The case study also demonstrates the potential benefit of including the engineer in the iterative procedure.

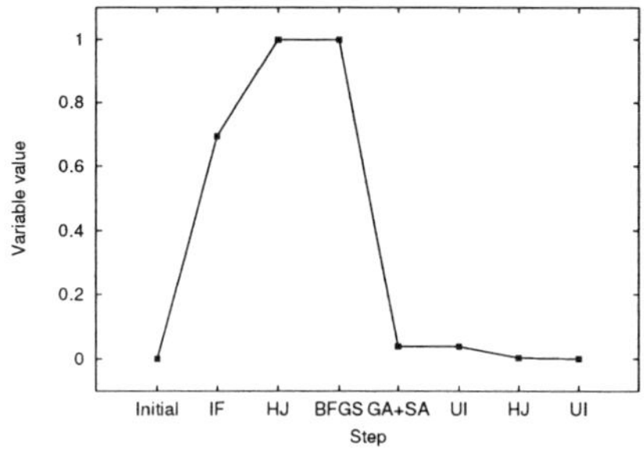

(a) x32

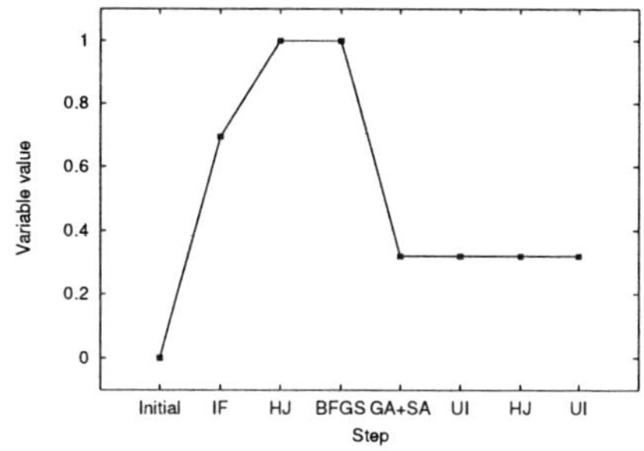

(b) sH31

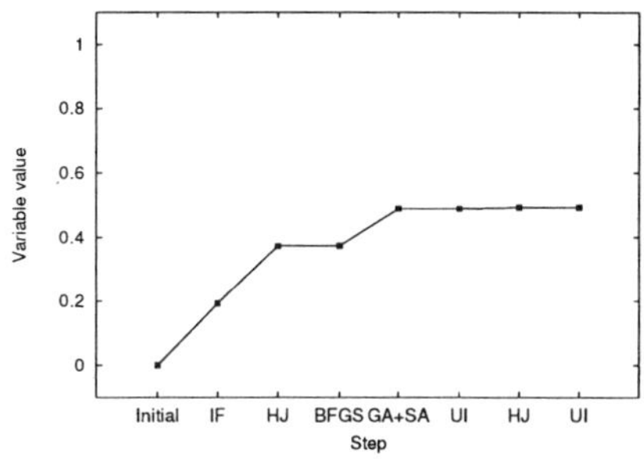

(c) sH12

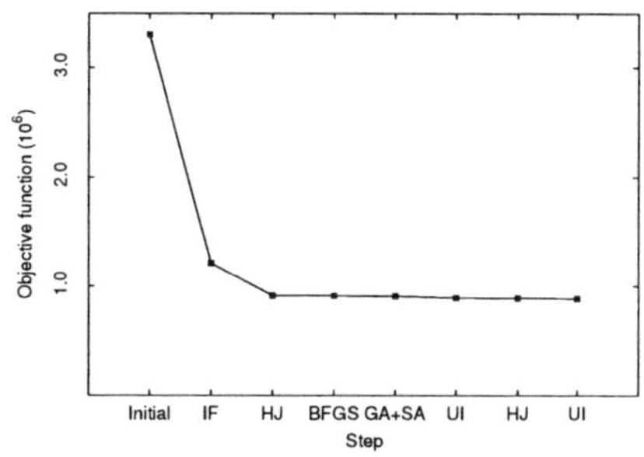

(d) $f(\underline{x})$

Figure 3. Evolution of a selection of decision variables and objective function for the case study.

References

1. J. Czyzyk, M. P. Mesnier and J. J. Moré, *IEEE Computing in Science and Engineering* **5**(3), 68–75 (1998).
2. J. E. Smith, In: P. M. Pardalos and H. E. Romeijn (eds.), *Handbook of Global Optimization Volume 2*, Kluwer Academic Publishers, 275–362 (2002).
3. P. J. M. van Laarhoven and E. H. L. Aarto, *Simulated Annealing: Theory and applications.* Kluwer Academic Publishers, Dordrecht (1987).
4. Published on the Internet. `ftp://ftp.gnu.org/gnu/gcide/`.
5. Ahuja, R. K., Ö. Ergun, J. B. Orlin and A. P. Punnen, *Discrete Applied Mathematics* **123**(1-3), 75–102 (2002).
6. S. B. Shouraki and G. Haffari, *Lecture Notes in Computer Science* **2510**, 102–109 (2002).
7. S. Prestwich, *Annals of Operations Research* **115**(1-4), 51–72 (2002).

8. P. van Hentenryck and L. Michel, *Lecture Notes in Computer Science* **3524**, 380–395 (2005).
9. M. Locatelli and F. Schoen, *Computational Optimization and Applications* **26**(2), 173–190 (2003).
10. R. Thomsen, *Biosystems* **72**(1-2), 57–73 (2003).
11. `http://en.wikipedia.org/wiki/Lamarckism`.
12. K. Ganesh and M. Punniyamoorthy, *International Journal of Advanced Manufacturing Technology* **26**(1-2), 148–154 (2005).
13. S. Ponnambalam and M. Reddy, *The International Journal of Advanced Manufacturing Technology* **21**(2), 126–137 (2003).
14. D. C. Tulpan and H. H. Hoos, *Lecture Notes in Computer Science* **2671**, 418–433 (2003).
15. F. V. Theos, I. E. Lagaris and D. G. Papageorgiou, *Computer Physics Communications* **159**(1), 63–69 (2004).
16. W. L. Price, *Journal of Optimization Theory and Applications* **40**(3), 333–348 (1983).
17. J. M. Bernardo and A. F. M. Smith, *Bayesian Theory.* John Wiley (2000).
18. D. G. Papageorgiou, I. N. Demetropoulos and I. E. Lagaris, *Computer Physics Communications* **109**(2-3), 227–249 (1998).
19. K. Smyth, H. H. Hoos and T. Stützle, *Lecture Notes in Computer Science* **2671**, 129–144 (2003).
20. N. Jussien and O. Lhomme, *Artificial Intelligence* **138**(1), 21–45 (2002).
21. B. Meyer and A. Ernst, *Lecture Notes in Computer Science* **3172**, 166–177 (2004).
22. Z.-J. Lee and C.-Y. Lee, *Information Sciences* **173**(1-3), 155–167 (2005).
23. E. S. Fraga and A. Žilinskas, *Advances in Engineering Software* **34**, 73–86 (2003).
24. E. C. Laskari, K. E. Parsopoulos and M. N. Vrahatis, *Numerical Algorithms* **34**(2-4), 393–403 (2003).
25. W. Xia and Z. Wu, *International Journal of Advanced Manufacturing Technology*, in press (2006).
26. E. S. Fraga, L. G. Papageorgiou and R. Sharma, In: A. Kraslawski and I. Turunen (eds.), *European Symposium on Computer Aided Process Engineering – 13*, Elsevier Science B.V., Amsterdam, *Computer-Aided Chemical Engineering* **14**, 119–124 (2003).
27. J. Aaltola, *Applied Thermal Engineering* **22**(8), 907–918 (2002).
28. G. F. Wei, P. J. Yao, X. Luo and W. Roetzel, *Chinese Journal of Chemical Engineering* **12**(1), 66–77 (2004).
29. G. W. A. Rowe and E. S. Fraga, In: I. C. Parmee (ed.), *Proceedings of Adaptive Computing in Design and Manufacture, ACDM'2006*, in press (2006).
30. W. Morton, *Proc. Inst. Mech. Eng. Part E* **216**(2), 89–104 (2002).
31. C. T. Kelley, *Iterative Methods for Optimization.* SIAM (1999).

AN MILP MODEL FOR MULTI-CLASS DATA CLASSIFICATION

G. XU AND L. G. PAPAGEORGIOU*

Centre for Process Systems Engineering, Department of Chemical Engineering, University College London, London WC1E 7JE, U.K.

This paper presents a multi-class data classification approach based on hyper-boxes using a mixed integer linear programming (MILP) model. Comparing with other discriminant classifiers, hyper-boxes are adopted to capture the disjoint regions and define the boundaries of each class so as to minimise the total misclassified samples. Non-overlapping constraints are specified to avoid overlapping of boxes that belong to different classes. In order to improve the training and testing accuracy, an iterative solution approach is presented to assign multi-boxes to single class. Finally, the applicability of the proposed approach is demonstrated through two illustrative examples from machine learning databases. According to the computational results, our approach is competitive in terms of prediction accuracy when comparing with various standard classifiers.

Keywords: Data classification, MILP, Hyper-boxes, Multiple classes

1. Introduction

Data classification is one of the main challenges in statistical pattern recognition. It deals with the identification of patterns through a series of training process and the assignment of new samples into known groups. In the last two decades, data classification has become an open question to the research community and various approaches such as neural networks (NN), support vector machines (SVM) and mathematical programming (MP) have been applied to design linear and nonlinear classifiers.

Mathematical programming (MP) techniques have been considered as one of the most powerful tools to develop classification models. Comparing with other approaches, MP methods are straightforward to implement through standard modeling tools and very few parameters are required during training processes. In recent years, most advances are stimulated by the work of Freed and Glover.[1] Most of these methods are linear programming (LP), which optimise deviations of misclassifications such as the maximisation of the

* To whom correspondence should be addressed: E-mail: l.papageorgiou@ucl.ac.uk

minimum deviation (MMD) or the minimisation of the sum of deviations (MSD). Some early attempts are reported by Erengue and Koehler.[2] Based on these LP models, MILP models are then derived to minimise the total number of misclassifications using various classifiers (see Fig. 1 for comparison). For example, Gehrlein[3] developed a successful MILP model to obtain linear discriminant functions which classify the data into two or more groups and maximise the number of correct classifications. Based on Gehrlein's work, Wilson[4] proposed an alternative MILP model and a hierarchical solution algorithm to overcome the computational difficulties. Recently, Sueyoshi[5-8] has addressed a series of non-parametric discriminant analysis approaches called DEA-DA (data envelopment analysis-discriminant analysis) from two-class to multi-class data classification problems. DEA-DA models provide a set of parallel linear discriminant functions to determine group memberships. Glen[9] presented an MILP approach so as to maximise the classification accuracy. Later, Glen[10] applied piecewise linear classifiers to approximate nonlinear discriminant functions to improve the classification performance.

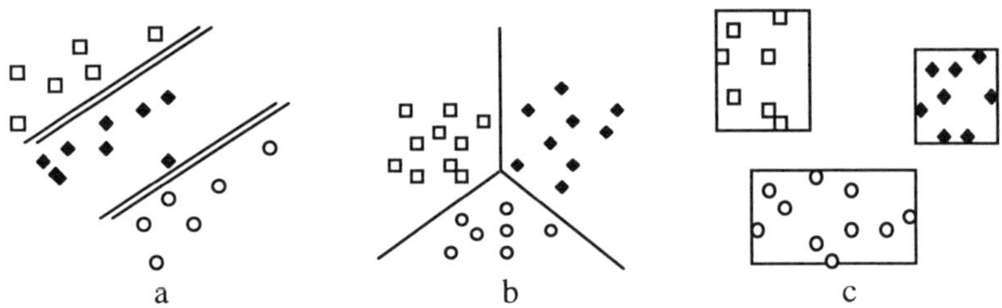

Figure 1. Discriminant classifiers: a. Parallel hyper-planes; b. Linear classifiers; c. Hyper-boxes.

Recently, hyper-boxes have become alternative classifiers to identify the boundaries of each pattern. Simpson[11] described a min-max neural networks classifier using fuzzy sets as pattern classes. N-dimensional fuzzy set hyper-boxes were defined by a minimum point and a maximum point with a corresponding membership function. The boundary points are determined through the fuzzy min-max learning algorithms. Mandal[12] addressed an efficient fuzzy partition of a feature space to generate fuzzy if-then rules for pattern classification. Overlapping hyper-boxes are used to decompose the whole feature space. Gabrys and Bargiela[13] generalized and extended Simpson's work to combine the supervised and unsupervised learning with a single learning algorithm. The boundaries of each distinct class are represented by hyper-boxes and the sizes of them will be adjusted through the process of learning. Finally, Succi and Pedrycz[14] proposed a two-phase hyper-box design process for classification problems. Several seed hyper-boxes are generated through clustering techniques and these candidate hyper-boxes are expanded through genetic algorithms.

2. Problem Statement and Mathematical Formulation

Consider a multi-class data classification problem with N classes and S samples. Each sample is characterized by M independent attributes. The class membership of each sample is known. In this paper, an MILP model for multi-class data classification problems is proposed so as to minimise the total number of misclassifications. Hyper-boxes are applied to enclose training data and non-overlapping constraints are introduced to discriminate data which belong to different classes.

The overall problem investigated can be stated as follows:

Given: Training data which are classified into N classes

Determine: The optimal position and dimensions of non-overlapping hyper-boxes

so as to Minimise the total misclassified samples.

The prototype of our model is based on the MILP formulations of process plant layout problems proposed by Papageorgiou and Rotstein[15] where process facilities are simplified to rectangular boxes and the optimal non-overlapping layouts will be determined so as to minimise the total connection cost of the process flowsheet. Our approach is also based on an MILP representation. The specific patterns of training data are captured by hyper-boxes with M dimensions (where M is the number of attributes). Linear constraints are used to avoid overlapping among hyper-boxes from different classes. The final objective function is the total number of misclassified samples, which is minimised. It is noted that the proposed MILP model assigns only one hyper-box to each class. Multi-boxes solution algorithms will be introduced to increase the training and testing accuracy (see Sec. 4).

3. Testing Procedure

After training, some samples with unknown labels will be assigned to one of the existing boxes through the testing process. The detailed testing strategy is as follows:

The distance between sample s and box i on attribute m, $DIST_{sim}$, is defined to be:

$$DIST_{sim} = Max\ (0, A_{sm} - UB_{im}, LB_{im} - A_{sm}) \qquad (1)$$

LB_{im} and UB_{im} are the lower and upper bound of box i on attribute m, which are determined respectively from the proposed MILP model. So, the distance between testing sample s and hyper-box i, DSS_{si}, is given by:

$$DSS_{si} = \sqrt{\sum_{m=1}^{M} DIST_{sim}^{2}} \tag{2}$$

According to (1), $DIST_{sim}$ is considered as the distance between s and the closest bound of box i on attribute m. Figure 2 shows the actual calculation of DSS_{si} in two-dimensional space. According to the two equations above, if a sample lies within a box on all attributes (see case a in Fig. 2), the distance between the sample and the box is zero. If the data is within the box on some attributes, the distance between sample s and box i on these attributes is zero and the overall distance is calculated via equation 2 (see case b in Fig. 2). If a sample is outside a box totally, the distance is shown in case c of Fig. 2. After calculating DSS_{si} for each s and each i, sample s will be assigned to the nearest box.

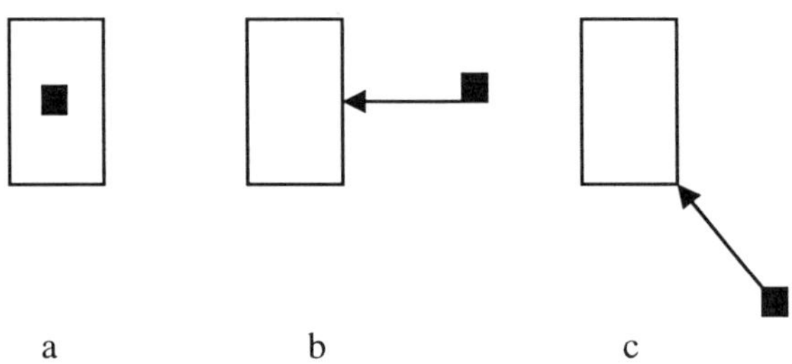

Figure 2. Calculation the distance between sample s and box i.

4. An Iterative Solution Algorithm

In this section, multi-boxes are introduced to improve the training and testing performance. After solving the single level MILP, new boxes are assigned to misclassified samples as determined by previous iterations. The algorithm will terminate when the objective function of two successive iterations keep the same value. The detailed algorithm is as follows:

STEP1: Initialize.
STEP2: Solve single level MILP (see section 2).
STEP3: Identify samples outside hyper-boxes (see section 3).
STEP4: Add one more box for each class with misclassified samples. Update the number of boxes for the training process.
STEP5. Solve single level MILP with the update hyper-boxes.
STEP6: If the objective function values of two successive iterations are the same, STOP; otherwise, go to STEP 3.

5. Computational Results

In this section, the proposed methodology is studied to evaluate the training and testing performance via two widely used machine learning datasets from UCI database (www.ics.uci.edu/mlearn/MLRepository.html). The first dataset used is the famous Fisher's iris data[16] which consists of 3 classes of 50 instances each. The second example consists of 178 wine recognition data samples which are the results of the chemical analysis of wines grown in the same region in Italy. Thirteen attributes are used to characterise each sample which belongs to one of three wine classes.

Training and testing evaluations will be carried out through the following schemes.

Scenario A: 50 percent of the samples in each class are extracted randomly for training and the rest are used for testing.

Scenario B: 50 percent of the whole samples are selected randomly and the rest samples are applied for testing

Scenario C: leave-one-out scheme.

The computational results from the iterative MILP are to be compared with two other MILP formulations[3,8] in terms of three different scenarios. We name the following models as follows:

Model_1_single: our single level MILP model;

Model_1_iter: our iterative MILP model;

Model_2: MILP model;[3] and

Model_3: MILP model.[8]

The proposed mathematical model and the iterative algorithm are implemented using the GAMS modeling system[17] with CPLEX mixed-integer programming solver with the margin of optimality of 1%. Scenario A and B are repeated 50 times and the mean prediction accuracy is reported.

Table 1. Comparative results.

Data	Model	Scenario A	Scenario B	Scenario C
iris	Model_1_single	94.98%	93.65%	92%
	Model_1_iter	**95.95%**	**94.02%**	**96%**
	Model_2	93.17%	92.64%	94%
	Model_3	93.94%	93.36%	94.67%
wine	Model_1_single	91.46%	90.22%	90.45%
	Model_1_iter	**91.68%**	91.42%	**94.38%**
	Model_2	91.07%	**91.78%**	92.69%
	Model_3	88.18%	88.76%	89.3%

The computational results for the two examples and three different scenarios investigated are shown in Table 1. The best performance (in terms of prediction accuracy) for each scenario is indicated in bold. It can clearly be seen that the iterative MILP (Model_1_iter) outperforms other approaches for most of the cases. Not surprisingly, the iterative MILP model performs better than the single level MILP since more than one box covers the same pattern.

6. Conclusions

In this paper, a rigorous mathematical model for multi-class classification problems has been proposed. According to the MILP representation, the optimal locations and dimensions of the hyper boxes are determined so as to minimise the total misclassifications. In the second stage, an iterative solution algorithm has been addressed to allow assignment of multi-boxes to one class. The applicability of the methodology has been demonstrated through two illustrative examples from machine learning databases. According to the computational results, the proposed method is competitive comparing with other alternative optimisation-based models.

References

1. N. Freed and F. Glover, *Decision Sci.* **12,** 68 (1981).
2. S. S. Erenguc and G. J. Koehler, *Manage. Decis. Econ.* **11**, 215 (1990).
3. W. V. Gehrlein, *Oper. Res. Let.* **5**, 299 (1986).
4. J. M. Wilson, *Int. J. Mgmt. Sci*, **24**, 681 (1996).
5. T. Sueyoshi, *Eur. J. Oper. Res.* ***115***, 564 (1999).
6. T. Sueyoshi, *Eur. J. Oper. Res.* **131**, 324 (2001).
7. T. Sueyoshi, *Eur. J. Oper. Res***.** **152**, 45 (2004).
8. T. Sueyoshi, *Eur. J. Oper. Res.* **169**, 247 (2006).
9. J. J. Glen, *J. Oper. Res. Soc.* **52**, 328 (2001).
10. J. J. Glen, *J. Oper. Res. Soc.* **56,** 331 (2005).
11. P. K. Simpson, *IEEE. T. Neural. Networ.* **3**, 776 (1992).
12. D. P. Mandal, *Pattern. Recogn.* **30**, 1971 (1997).
13. B. Gabrys and A. Bargiela, *IEEE. T. Neural Networ.* **11**, 769 (2000).
14. G. Succi and W. Pedrycz, *J. Syst. Software.* **76**, 277 (2005).
15. L. G. Papageorgiou and G. E. Rotstein, *Ind. Eng. Chem. Res.* **37**, 3631 (1998).
16. R. Fisher, *Ann. Eugenics* **7**, 179 (1936).
17. A. Brooke. D. Kendrick and R. Raman, *GAMS: A user's guide*. GAMS Development Corp., Washington, DC (1998).

IMPLEMENTATION OF PARALLEL OPTIMIZATION ALGORITHMS USING GENERALIZED BRANCH AND BOUND TEMPLATE*

M. BARAVYKAITĖ

Vilnius Gediminas Technical University, Saulėtekio al. 11
LT-10223 Vilnius, Lithuania, E-mail: mmb@fm.vtu.lt

J. ŽILINSKAS

Institute of Mathematics and Informatics, Akademijos 4
LT-08663 Vilnius, Lithuania, E-mail: julius.zilinskas@mii.lt

In this paper a template for implementation of parallel branch and bound algorithms is considered. Standard parts of branch and bound algorithms are implemented in the template and only method specific rules should be implemented by the user. The sequential version of the template allows easy testing of different variants of algorithms. Using parallel version of the template the user can obtain parallel programs without actually doing parallel programming. Several parallel global and combinatorial optimization algorithms have been implemented using template and the results are presented.

1. Introduction

Many problems in engineering, physics, economic and other fields may be formulated as optimization problems, where the minimum value of an objective function should be found.

Branch and bound (BB) is a general algorithm to solve optimization problems. Its general structure can be implemented as an algorithm template that will simplify implementation of specific BB algorithms to solve particular problem. Only specific parts of the algorithm should be implemented by the user. Same template ideas applied for parallel programming can relieve users from actual parallel programming.

*The authors wish to acknowledge the support of the HPC-Europa programme, funded under the European Commission's Research Infrastructures activity of the Structuring the European Research Area programme, contract number RII3-CT-2003-506079, and Lithuanian State Science and Studies Foundation.

2. General Branch and Bound Algorithm

Consider minimization problem, formulated as follows

$$f^* = \min_{X \in D} f(X). \tag{1}$$

where $f(X)$ is an objective function, X are decision variables, and D is a search space. Besides of the minimum f^*, one or all minimizers X^* : $f(X^*) = f^*$ should be found.

Branch and bound is a technique that can be used in many optimization algorithms, for example to solve combinatorial optimization or covering global optimization problems. The idea is to detect the subspaces not containing the global minimum and discard them from further search. According to BB algorithm an initial approximation of the problem solution should be initiated first. Initial search space is subsequently divided into smaller subspaces. Then each subspace is evaluated trying to find out if it can contain optimal solution. For this purpose a lower bound for objective function is calculated over the subspace and compared with upper bound for the minimum value. If the lower bound for objective function over the subspace is larger than the upper bound for minimum value, the subspace cannot contain the global minimizer and therefore it is rejected from further search. Otherwise it is inserted into the list of unexplored subspaces. The algorithm finalizes when there are no subspaces in the list. The general branch and bound algorithm is shown in Algorithm 1, where L denotes list of candidates, S denotes the solution, $UB(D_i)$ and $LB(D_i)$ denote upper and lower bounds for minimum value of the objective function over subspace D_i.

Cover solution space D by $L = \{L_j | D \subseteq \bigcup L_j, j = \overline{1,m}\}$ using **covering rule**.
$S = \emptyset, \quad UB(D) = \infty$.
while subspace list is not empty $L \neq \emptyset$ **do**
    Choose $I \in L$ using **selection rule**, exclude I from L.
    if $LB(I) < UB(D) + \epsilon$ **then**
        Branch I into p subspaces I_j using **branching rule**.
        for all $I_j, j = \overline{1,p}$ **do**
            Find $UB(I_j \bigcap D)$ and $LB(I_j)$ using **bounding rules**.
            $UB(D) = \min(UB(D), UB(I_j \bigcap D))$.
            if $LB(I_j) < UB(D) + \epsilon$ **then**
                if I_j can be a solution **then** $S = I_j$.
                else $L = \{L, I_j\}$.

Algorithm 1: General branch and bound algorithm.

Before the cycle of iterations the list of candidate subspaces should be initialized by covering search space by one or more subspaces. In combinatorial optimization subspace can be a solution if it is indivisible, in global optimization if it is a small sub-region bracketing a potential solution with predefined accuracy. The rules of covering, selection, branching and bounding differ from algorithm to algorithm.

Main strategies of selection of candidate subspaces are the following:

- **Best first.** Select a candidate with minimal lower bound. Candidate list L can be implemented using *heap* or *priority queue.*
- **Depth first.** Select the youngest candidate. First-In-Last-Out structure is used for candidate list which can be implemented using *stack.*
- **Breadth first.** Select the oldest candidate. First-In-First-Out structure is used for candidate list which can be implemented using *queue.*
- **Improved selection.** It is based on heuristic[9,3] or probabilistic[4] criteria. Candidate list can be implemented using *heap* or *priority queue.*

The candidate selection strategies influence the efficiency of branch and bound algorithm and the number of candidates kept in the candidate list. For particular problems some strategies can considerably improve the performance of the algorithm.

The bounding rule describes how the bounds for minimum of the objective function are found. For the upper bound for minimum over the search space $UB(D)$ the best currently found value of the objective function might be accepted.

3. BB Algorithm Template

3.1. *Template Programming*

The idea of the template programming is to implement general structure of the algorithm that could be later used to solve different problems. All general features of the algorithm and its interaction with the particular problem must be implemented in the template. The particular features related to the problem must be given by the template user. The user only has to identify the needed algorithm, choose the right template and implement problem dependent parts. Templates ease programming, clear algorithm logic, allow easy re-use of the implementation.

Template based programming can be very useful in parallel programming.[15,7] Parallel algorithm template must fully or partially specify the main features of a parallel algorithm: partitioning, communication, agglomeration and mapping. From the user's point of view, all or nearly all the coding should be sequential; all, or almost all the parallel aspects should be provided by the tool.

Often parallel programs are created by parallelizing the existing sequential programs. Then parallel algorithm template can use features implemented by the sequential algorithm template. If a sequential template was used to create the sequential program, then there is no need to rewrite existing code to obtain parallel one. In this way, templates save time and efforts of the users. On the other hand generalization of main parallel aspects of algorithms may result in lower efficiency of the implementation. Some examples of parallel templates of different algorithms are MST,[13] Mallba,[1] CODE.[15] Some examples of BB parallelization tools are BOB,[11] PICO,[5] PPBB,[16] PUBB.[14]

3.2. *Implementation of BB Template*

We present a template implementation of general BB algorithm.[2] BB algorithm template is implemented using C++ objected oriented paradigm and inheritance. MPI[12] is used for underlying communications. Algorithm class scheme is presented in Fig. 1.

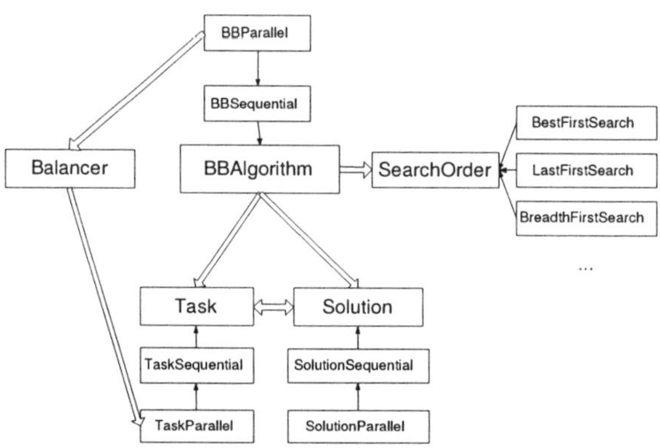

Figure 1. Template class scheme.

`BBAlgorithm` implements various sequential and parallel BB algorithms. The algorithm is performed using `Task`, `Solution` and `SearchOrder` instances. The implementation of the `BBAlgorithm` is given in the template but the user can extend this class.

`SearchOrder` defines the strategies how to select next subspace from the list of subspaces for subsequent partitioning. The most popular strategies are already implemented as methods and they are ready for application. The user can implement his/her own specific rules, in this case he/she should define methods `Insert`, `Delete`, `QueueSize`, `QueueEmpty`.

Class `Task` defines the problem to be solved. It should implement the basic BB algorithm methods: `Initialize`, `Branch`, `Bound`. Some often used `Branch` methods are implemented in the template. Standard `Bound` calculation methods such as for Lipschitz functions are included in the template as well.

Class `Solution` implements the solution to be found and should be implemented by the user.

Class `Balancer` is used for parallel applications to balance the processor load.

The user has to fill in the particular `Task` and `Solution` class instances and compile the selected variant of the program. The template can be extended with useful methods and algorithms.

Sequential usage of BB template. When used for sequential programming, the tool allows to reuse once implemented problem specific parts of the algorithm to test different variants of BB algorithm and search strategies. As an example of combinatorial optimization the solution of traveling salesman problem (TSP) over 20 cities[10] has been implemented. As an example of global optimization, Lipschitz optimization of function described in the chapter 'Lipschitz Optimization' in[8] has been implemented. Both problems with different search orders have been tested and a result has been obtained that for TSP it takes 50851 tasks to solve the problem using best first search, 99278 using last first search and 348990 using breadth first search. For Lipschitz function best first search took 788 tasks, depth first search 3327, breadth first search 694053.

Parallel usage of BB template. Any parallel algorithm for a given problem attempts to divide it into sub-problems which can be solved concurrently on different processors. Four main steps are performed during development of a parallel algorithm:[6] partitioning, communication, agglomeration, mapping.

The aim of *partitioning* is to decompose the computations into subtasks. Attention is focused on recognizing opportunities for parallel execution. During this step we should take into account that a larger number of subtasks gives more possibilities to improve a load balancing among processors, but at the same time it increases data communication costs. Then *communications* required to coordinate task execution is determined. In *agglomeration* step, if necessary, tasks are combined into larger tasks to improve performance and to reduce development and communication costs. This step is not necessary for parallel BB algorithms implemented in the template. Then each subtask is *mapped* or assigned to a processor. During this step we try to minimize the computation time by preserving a good load balance among processors.

If the BB template is used for sequential programming, then in order to get a parallel variant of the program, the user has to select one of proposed parallel algorithms and compile parallel version of the program. There are several implemented parallel algorithms in the template.

In the first algorithm, initial search space is divided into several large subspaces that are mapped to the processors and the algorithm presented in Algorithm 1 is performed independently on each processor. The number of subspaces coincides with the number of processors. We will call it a parallel BB algorithm with a static distribution of job pool (SJP).

The search space can be divided into M subspaces, where M is much larger than the number of processors. Then subspaces are distributed *a priori* among processors in random order. This algorithm is called RJP parallel BB algorithm with a random distribution of job pool.

A subspace is eliminated from the further search by comparing the lower bound for objective function over the subspace with the upper bound $UB(D)$. The best currently found value of objective function can be used for the upper bound. In previously described parallel algorithms processors know only values of objective function found in the subspaces mapped to the particular processor. In some situations this can result in slower subspace elimination. Processors can share $UB(D)$. When new value of the upper bound is found, it is broadcasted to the other processors. In order not to stop calculations, this exchange is performed asynchronously. These modifications of the BB algorithm will be called SJP SE and RJP SE, depending on the rule to distribute the initial job pool.

Calculation experiments were performed on up to 15 nodes of Vilnius Gediminas Technical university computer cluster *Vilkas (www.vilkas.vtu.lt)* and up to 256 nodes of *IDRIS* supercomputer *(www.idris.fr)*. Figures 2 and 3 present the calculation times.

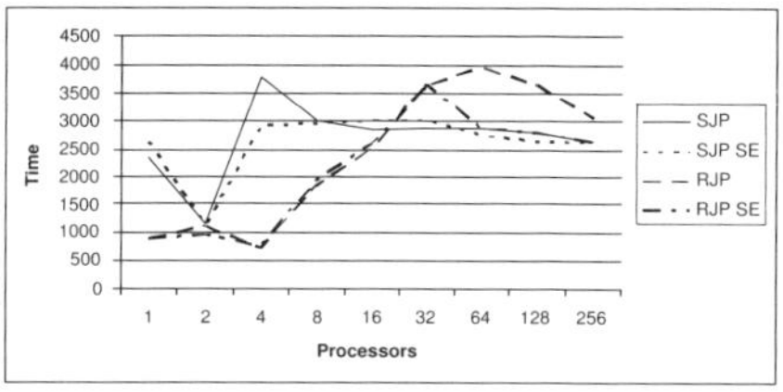

Figure 2. Parallel execution time of TSP using best first search.

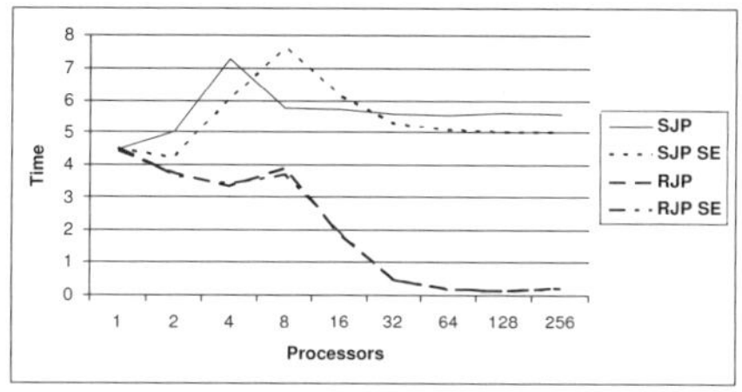

Figure 3. Parallel execution time of Lipschitz function using best first search.

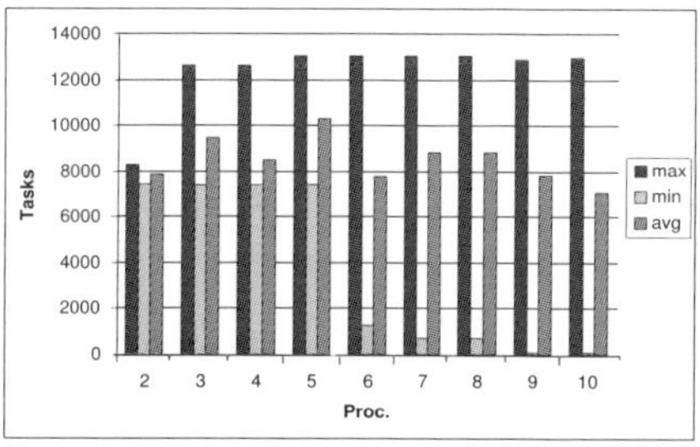

Figure 4. Disbalance of Lipschitz function using the best first search.

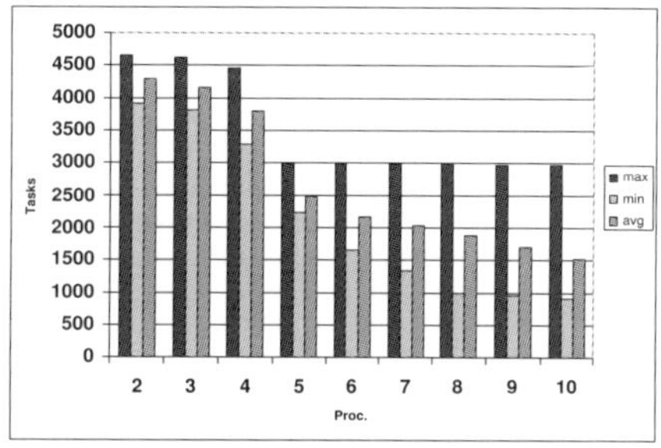

Figure 5. Disbalance of Lipschitz function using the best first search and load balancing.

To obtain better parallel execution, BB template is improved with balancing module. The module tries to equalize the load by passing some tasks from heavily loaded processor to the idle one. Figures 4 and 5 present processor load disbalance without and with dynamic load balancing. Here *"max"*, *"min" and* "avg" shows maximum, minimum and average number of tasks executed among the processors. For this particular case, load balancing reduces both the difference between maximum and minimum tasks and total amount of executed tasks.

4. Conclusions

In this work the use of the template for implementation of branch and bound algorithms is presented. The main aim of this tool is to ease the implementation of sequential and parallel programs for combinatorial optimization and covering methods for global optimization. Standard parts of branch and bound algorithm are implemented in the template and only method specific rules should be implemented by the user. Parallel programs can be obtained automatically using sequential program implemented using the template.

References

1. E. Alba, F. Almeida *et all.*, *MALLBA: A Library of Skeletons for Combinatorical Optimization*, Departamento de Lengua jes y Ciencias de la Computacion Universidad de Malaga, ALCOMFT-TR-02-120, (2001).
2. M. Baravykaitė, R. Čiegis and J. Žilinskas, Template realization of generalized branch and bound algorithm, *Matematical Modelling and Analysis* **10**(3), 217–236 (2005).
3. T. Csendes, Generalized subinterval selection criteria for interval global optimization, *Numerical Algorithms* **37**, 93–100 (2004).
4. M. Dür and V. Stix, Probabilistic subproblem selection in branch-and-bound algorithms, *Journal of Computational and Applied Mathematics* **182**, 67–80 (1980).
5. J. Eckstein, C. A. Phillips and W. E. Hart, *PICO: An Object-Oriented Framework for Parallel Branch and Bound*, RUTCOR Research Report, Rutgers University, Piscataway, NJ 40-2000 (2000).
6. I. Foster, *Designing and Building Parallel Programs*, Addison-Wesley, (1995).
7. D. Goswami, A. Singh and B. Preiss, From design patterns to parallel architecture skeletons, *jpdc* **62**(4), 669–695 (2002).
8. R. Horst, P. M. Pardalos and N. V. Thoai, *Introduction to Global Optimization, 2nd edition,* Nonconvex Optimization and its Applications **48**, Kluwer Academic Publishers (2001).
9. V. Kreinovich and T. Csendes, Theoretical justification of a heuristic subbox selection criterion for interval global optimization, *Central European Journal of Operations Research*, **9**(3), 255–265 (2001).
10. E. W. Lawler, J. K. Lenstra, A. Rinnooy Kan and D. B. Smoys, *The Traveling Salesman Problem : A Guided Tour of Combinatorial Optimization*, Wiley Series in Discrete Mathematics and Optimization, John Wiley & Sons, (1985).
11. B. Le Cun and C. Roucairol, *BOB: a Unified Platform for Implementing Branch-and-Bound like Algorithms*, Universite de Versailles - Laboratoire PRiSM (1995).
12. Message Passing Interface Forum, *MPI: A Message-Passing Interface Standard (Version 1.1)* (1995).
13. R. Šablinskas, *Investigation of algorithms for distributed memory parallel computers*, Vytautas Magnus University, (1999).
14. Y. Shianno and T. Fujier, *PUBB (Parallelization Utility for Branch-and-Bound algorithms) User Manual*, Version 1.0. (1999).
15. A. Singh, J. Schaeffer and D. Szafron, Views on template-based parallel programming, *CASCON 96 CDRom Proceedings*, Toronto, October (1996).
16. S. Tschoke and T. Polzer, *Portable Parallel Branch-and-Bound Library PPBB-Lib. User Manual, Version 2.0.*, Department of Computer Science, University of Paderborn (1996).

APPLICATION OF STOCHASTIC APPROXIMATION IN TECHNICAL DESIGN

V. BARTKUTĖ AND L. SAKALAUSKAS

Institute of Mathematics and Informatics, Akademijos st. 4
Vilnius 08663, Lithuania

In this paper, we consider problems related to the implementation of Stochastic Approximation (SA) in technical design, namely, estimation of a stochastic gradient, improvement of convergence, stopping criteria of the algorithm, *etc*. The accuracy of solution and the termination of the algorithm are considered in a statistical way. We build a method for estimation of confidence interval of the objective function extremum and stopping of the algorithm according to order statistics of objective function values provided during optimization. We give some illustration examples of application of developed approach of SA to the optimal engineering design problems, too.

1. Introduction

In many practical problems of technical design some of the data may be subject to significant uncertainty which is reduced to stochastic-statistical models. Models applied in such cases are appropriate when data evolve over time and decisions need to be made prior to observing the entire data stream. Consequently, the performance of such problems can also be viewed like constrained stochastic optimization programming tasks. SA can be considered as alternative to traditional optimization methods, especially when objective functions are no differentiable or computed with noise.

Application of SA to nonsmooth optimization is both a theoretical and practical problem. Computational properties of SA algorithms are mainly determined by the approximation approach to the stochastic gradient.[1-6] At last time Simultaneous Perturbation Stochastic Approximation methods (SPSA) become rather popular in literature devoted to stochastic search. It is of interest to consider SPSA methods because in these methods values of the function for estimating the stochastic gradient are required only at one or several points. The SPSA algorithms were considered by several authors who used various smoothing operators. SPSA methods, uniformly smoothing the function in an n-dimensional hypercube, are described.[4] SPSA algorithm with the Bernoulli perturbation was proposed and the computational efficiency of SPSA as compared with the standard finite difference approximation was indicated.[5] The convergence and asymptotic behaviour of this algorithm were established in the class of differentiable functions.

Application of the SPSA algorithms to non-differentiable optimization is of particular theoretical and practical interest. In this paper, we focus on the objective functions from the Lipschitz class. We consider and compare SPSA algorithms with various perturbation operators and SA finite difference algorithms. In this paper, we consider problems related to the implementation of SA algorithms, for example, estimation of a stochastic gradient, improvement of convergence, stopping criteria of the algorithm, *etc.* We build a method for estimation of confidence interval of the objective function extremum and stopping of the algorithm according to order statistics of objective function values provided during optimization. We give some illustration examples of applications of this method to the optimal engineering design problems, too.

2. Formulation of the Optimization Problem

Let optimization problem is

$$f(x) \to \min, \tag{1}$$

where the objective function $f : \Re^n \to \Re$ is Lipschitzian. Let $\partial f(x)$ be the generalized gradient (GG), i.e., the Clarke subdifferential[7] of this function. For the solving of the problem (1) we consider and compare three stochastic approximation methods: SPSA with Lipschitz perturbation operator, SPSA with Uniform perturbation operator and Standard Finite Difference Approximation (SFDA) method. General scheme of SA approach is as follows

$$x^{k+1} = x^k - \rho_k \cdot g^k, \quad k = 1, 2, \dots, \tag{2}$$

where g^k is the value of the stochastic gradient estimator at the point x^k, ρ_k is a scalar multiplier in iteration k. This scheme is the same for different SA algorithms whose distinguish only by approach for stochastic gradient estimation:

A) gradient estimator of the SPSA with Lipschitz perturbation operator is expressed as:

$$g(x, \sigma, \xi) = \frac{(f(x + \sigma\xi) - f(x)) \cdot \xi}{\sigma \cdot \|\xi\|} \tag{3}$$

where σ is the value of the perturbation parameter, vector ξ is uniformly distributed in the unit ball $\psi(y) = \begin{cases} \frac{1}{V_n}, & \text{if } \|y\| \le 1, \\ 0, & \text{if } \|y\| > 1. \end{cases}$, V_n is the volume of the *n*-dimensional ball;[8]

B) the same estimator for the SPSA with Uniform perturbation operator is expressed as:

$$g(x, \sigma, \xi) = \frac{(f(x + \sigma \cdot \xi) - f(x - \sigma \cdot \xi)) \cdot \xi}{2\sigma}, \tag{4}$$

where σ is the value of the perturbation parameter, $\xi = (\xi_1, \xi_2, \ldots, \xi_n)$ is a vector consisting of variables uniformly distributed from the interval [-1;1];[4]

C) gradient estimator $g = (g_1, g_2, \ldots, g_n)$ for SFDA has following components:

$$g_i(x, \sigma, \xi, \upsilon) = \frac{f(x + \sigma \cdot \xi + \upsilon \cdot \varepsilon_i) - f(x + \sigma \cdot \xi)}{\upsilon}, \tag{5}$$

where ξ is the same like in (3), $\varepsilon_i = (0,0,0,\ldots,1,\ldots,0)$ is the vector with zero components except i^{th} one, which is equal to 1, υ and σ are the values of the perturbation parameters.[4]

Note, that only two function values have to be computed for the estimation of the stochastic gradient according to SPSA methods A) and B), and the SFDA method, in its turn, requires that *n+1* function values be computed.

Method (2) converges a. s. for all considered gradient estimators under certain conditions.[8,4] The SPSA convergence rate

$$E\left\|x^k - x^*\right\|^2 = O\left(\frac{1}{k^\gamma}\right), 1 \le \gamma < 2 \tag{6}$$

has been established in a presence of function computed without noise.[8]

3. Computer Modelling

The convergence of the proposed method was studied by computer modelling, as well. We considered a class of test function $f = \sum_{k=1}^{n} a_k |x_k|^\beta + M$, where a_k were randomly and uniformly generated in the interval $[\mu, K]$, $K > \mu > 0$. The samples of T=500 test functions were generated, when $\mu = 2$, *K*=5.

Table 1. Empirical and theoretical rates of convergence by SA methods.

	$\beta = 0.5$	$\beta = 0.75$	$\beta = 0.9$	
Theoretical	1.5	1.75	1.9	
Empirical				
SPSA (Lipshitz perturbation)				
$n = 2$	1.45509	1.72013	1.892668	$\rho_i = n \cdot \min\left(0.005; \frac{0.25}{i}\right)$
$n = 4$	1.41801	1.74426	1.958998	$\sigma_i = \sqrt{\frac{(n+2)(n+3)}{n(n+1)}} \min\left(0.01; \frac{1}{i^{\beta}}\right)$
SPSA (Uniform perturbation)				
$n = 2$	1.605244	1.938319	1.988265	$\rho_i = n \cdot \min\left(0.055; \frac{1}{i}\right)$
$n = 4$	1.551486	1.784519	1.998132	$\sigma_i = \sqrt{\frac{(n+2)(n+3)}{n(n+1)}} \min\left(0.05; \frac{0.5}{i^{\beta}}\right)$
Stochastic Difference Approximation method				
$n = 2$	1.52799	1.76399	1.90479	$\rho_i = \min\left(0.5; \frac{0.25}{i}\right)$
$n = 4$	1.50236	1.75057	1.90621	$\sigma_i = \sqrt{\frac{(n+2)(n+3)}{n(n+1)}} \min\left(0.1; \frac{1}{i^{\beta}}\right)$; $\upsilon_i = \frac{\sigma_i}{\sqrt{i}}$

The test functions were minimized by algorithms described in Sec. 2 and rate of convergence $\Delta_k = E\left\| x^k - x^* \right\|^2$ was investigated. The empirical least square estimates of the rate of convergence by the Monte-Carlo method and coefficients of optimizing sequence (2) are presented in Table 1. Theoretical rates of convergence are given for comparison, too. Hence, empirical rates of convergence look similarly for all methods and corroborate theoretical conclusions (6).

4. Application of Order Statistic to Optimality Testing

Application of extreme value theory in optimization algorithms was studied by several authors, however this interesting topic is not studied sufficiently well. Statistical inferences about the maximal (minimal) value of a function are described.[9-11] An application of the extreme order statistics to the estimation of the location of the maximum of a regression function can be found.[12]

Let consider description of the inference of order statistics to optimality testing. Assume, the problem (1) be solved by the SA algorithms providing a sample

$$H = \{\eta_1, \ldots, \eta_N\}, \tag{7}$$

whose elements are function values $\eta_k = f(x_k)$ provided during optimization. We build a method for the estimation of the minimum of the objective function by order statistics of sequence (7). We will apply for this purpose the extreme value theory of i. i. d. variables and examine our approach by experimental way, because theoretical distribution of extremes of sequence (7) is not studied yet.

Thus, to estimate confidence intervals for the minimum *A* of the objective function, it suffices to choose from sample H only *m+1* order statistic: $\eta_{(0)}, \ldots, \eta_{(m)}$, where $m = m(N)$, $\frac{m^2}{N} \to 0,\ N \to +\infty$.[9,10] Then the linear estimators for *A* can be as follows:

$$A_{N,m} = \sum_{i=0}^{m} a_i \eta_{(i)}, \tag{8}$$

where *m* is much smaller than *N*, $a_0, \ldots, a_m$ are some coefficients satisfying the condition $\sum_{i=1}^{m} a_i = 1$. Let us consider simple sets of coefficients:[13] $a = (1 + c_m,\ 0,\ \ldots,\ 0,\ -c_m)$, $c_m = \frac{1}{\prod_{i=1}^{m}\left(1 + \frac{1}{i \cdot \alpha}\right) - 1}$, where α is the parameter of extreme values distribution . We examine a choice of this parameter for continuous optimization

$$\alpha - \frac{n}{\varphi}, \tag{9}$$

where φ is the parameter of homogeneity of the function $f(x)$ in the neighbourhood of the point of minimum.[10]

The one-side confidence interval of the minimum of the objective function is as follows:

$$[\,\eta_{(0)} - r_{m,\gamma} \cdot (\eta_{(m)} - \eta_{(0)}),\ \eta_{(0)}\,], \tag{10}$$

where $r_{m,\gamma}$ - certain constant, γ is the confidence level.[10]

We investigate the approach developed by computer modeling for SPSA method with Lipschitz perturbation, where parameters taken from the Table 1,

$\gamma = 0.95$. The upper and lower bounds of the confidence interval for the minimal value of the objective function are given in Table 2 and Fig. 1. These bounds were estimated by the Monte-Carlo method with the number of iterations N=10000 and the number of trials T=500, when the confidence level was $\gamma = 0.95$. From Fig. 1 we can see that the length of the confidence interval decreases when the number of iterations increases.

Table 2. Experimental results of the one-side confidence interval.

N=10000, T=500, $\gamma = 0.95$	Estimate of $A = 0$	Confidence interval		Empirical probability of hitting of *A* to the confidence interval
		Lower bound	Upper bound	
n=2	0.000014	-0.000036	0.0000275	0.954
n=4	0.000289	-0.000569	0.0005355	0.946
n=6	0.001027	-0.001879	0.001819	0.942
n=10	0.004488	-0.005359	0.0070606	0.946

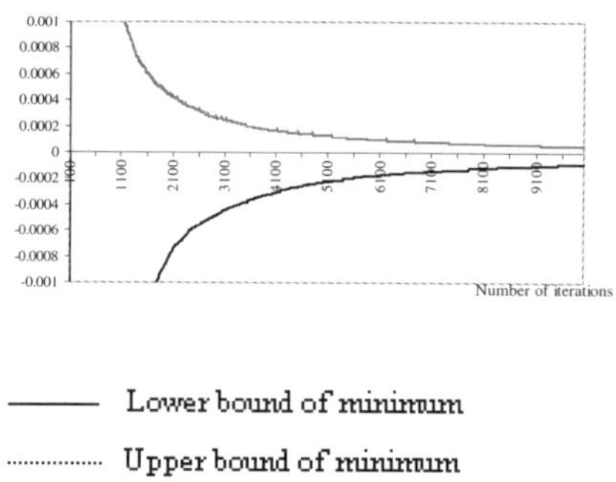

Figure 1. Confidence bounds of the minimum (A=0, $\gamma = 0.95$, $n = 2$, $T = 500$).

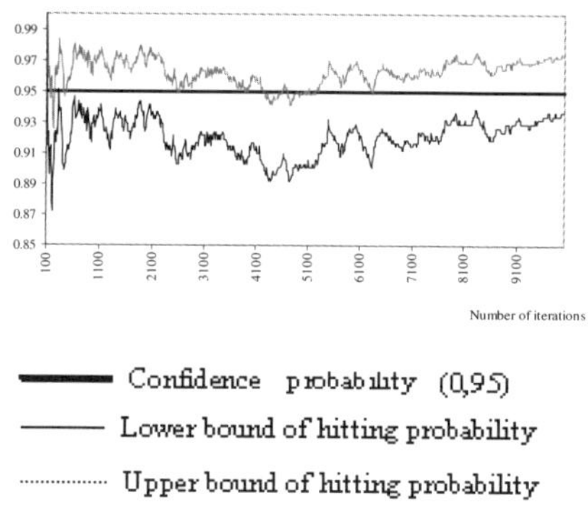

Figure 2. Confidence interval of the hitting probability $n = 2$, $\gamma = 0.95$, $T = 500$).

Thus, from the results of Table 2 and Fig. 1 we can see that formula (10) approximates the confidence interval of objective function minimum rather well. Results of Table 2 and Fig. 2 show also that empirical probability of minimal value hitting the confidence interval corroborates well to theoretical admissible confidence level γ. From the experimental results it follows that formula (10) can be used to create the stopping criteria for the algorithm, namely, the algorithm stops when the length of the confidence interval becomes smaller than the admissible value $\varepsilon > 0$.

5. Computer Modelling

We will demonstrate the applicability of SA for two real-life problems.

5.1. *Volatility Estimation by Stochastic Approximation Algorithm*

Financial engineering, as well as risk analysis in the market research and management is often related to the implied and realized volatility. Let us consider the application of SA to the minimization of the mean absolute pricing error for the parameter calibration in the Heston stochastic volatility model.[14] In this model option pricing biases can be compared to the observed market prices, based on the latter solution and pricing error. We consider the mean absolute pricing error (MAE) defined as:

$$\mathrm{MAE}(\kappa, \sigma, \rho, \lambda, v, \theta) = \frac{1}{N}\sum_{i=1}^{N}\left|C_i^H(\kappa, \sigma, \rho, \lambda, v, \theta) - C_i\right|, \tag{11}$$

where N is the total number of options, C_i and C_i^H represent the realized market price and the implied theoretical model price, respectively, while $\kappa, \sigma, \rho, \lambda, v, \theta$ (n=6) are the parameters of the Heston model to be estimated.

To compute option prices by the Heston model, one needs input parameters that can hardly be found from the market data. We need to estimate the above parameters by an appropriate calibration procedure. The estimates of the Heston model parameters are obtained by minimizing MAE:

$$\mathrm{MAE}(\kappa, \sigma, \rho, \lambda, v, \theta) \to \min . \tag{12}$$

Heston model was implemented for the Call option on SPX (29 May 2002). The SPSA algorithm with Lipschitz perturbation was applied to the calibration of the Heston model. Usually, SPSA requires that MAE be computed several hundred times that is reasonable for interactive Heston model calibration. Figure 3 below illustrates the applicability of the SPSA algorithm in practice, where we can see the dependence of MAE on the number of computations of function values compared with the same dependence obtained by the SFDA method.

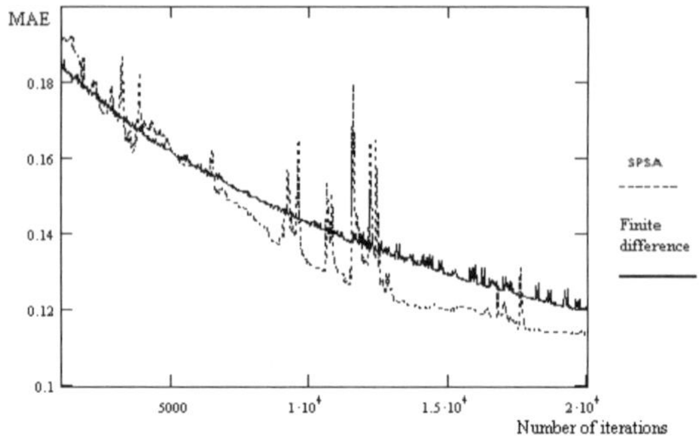

Figure 3. Minimization of MAE by SPSA and SFDA methods.

5.2. *Optimal Design of Cargo Oil Tankers*

In cargo oil tankers design, it is necessary to choose such sizes for bulkheads, that the weight of bulkheads would be minimal. After some details the minimization of weight of bulkheads for the cargo oil tank we can formulate like nonlinear programming task:[15]

$$f(x) = \frac{5.885 \cdot x_4 (x_1 + x_3)}{x_1 + \sqrt{(x_3^2 - x_2^2)}} \to \min, \tag{13}$$

subject to $g_1(x) = x_2 x_4 \left(0.4 \cdot x_1 + \frac{1}{6} x_3\right) - 8.94 \cdot \left(x_1 + \sqrt{(x_3^2 - x_2^2)}\right) \geq 0$,

$$g_2(x) = x_2^2 x_4 \left(0.2 \cdot x_1 + \frac{1}{12} x_3\right) - 2.2 \cdot \left(8.94 \cdot \left(x_1 + \sqrt{(x_3^2 - x_2^2)}\right)\right)^{\frac{4}{3}} \geq 0,$$

$g_3(x) = x_4 - 0.0156 x_1 - 0.15 \geq 0$,

$g_4(x) = x_4 - 0.0156 x_3 - 0.15 \geq 0$,

$g_5(x) = x_4 - 1.05 \geq 0$,

$g_6(x) = x_3 - x_2 \geq 0$, where x_1 - width, x_2 -debt, x_3 - length, x_4 - thickness.

Let us consider the application of SA to the minimization of the bulkheads weight by the penalty method. In Fig. 4 the penalty function and the best feasible objective functions under the number of iterations minimized by SPSA with Uniform perturbation and SFDA method are depicted. In Fig. 5 the averaged upper and lower bounds of the minimum are illustrated. For comparison the function minimum value is presented.[15] As we see, the linear estimators by order statistics make it possible to evaluate minimum with admissible accuracy and introduce rule for algorithm stopping when the confidence interval becomes less than certain small value.

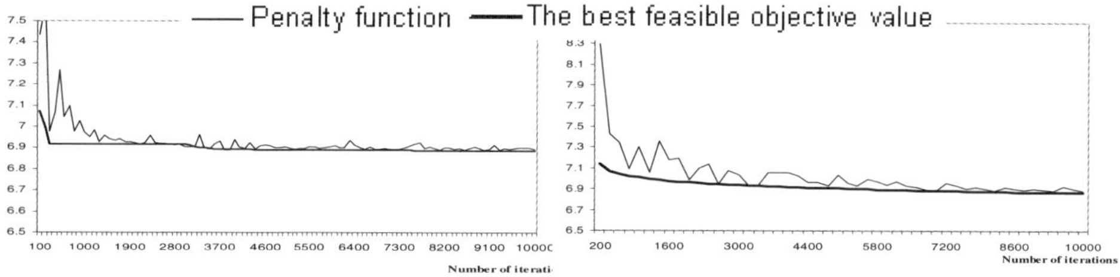

Figure 4. SPSA with Lipschitz perturbation and SFDA for the cargo oil target design.

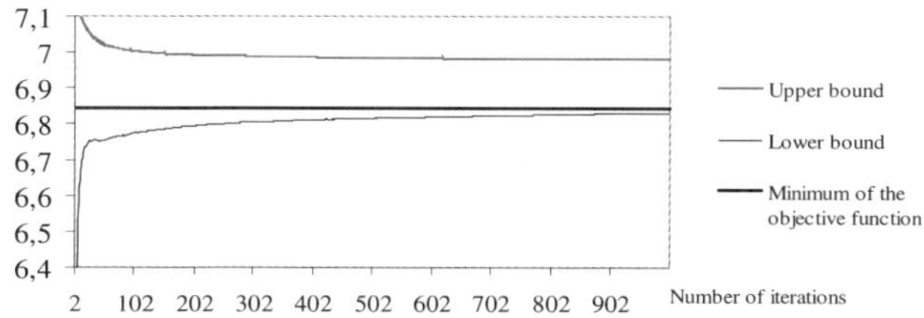

Figure 5. Confidence bounds of the minimum (A=6.84241, T=100, N=1000).

6. Conclusion

Application of SA to engineering problems has been considered comparing three algorithms. The rate of convergence of the developed approach was explored for the functions with a sharp minimum by computer simulation when there are no noises in computation of the objective function. Computer simulation by Monte-Carlo method has shown that the empirical estimates of the rate of convergence corroborate the theoretical estimation of the convergence order $O\left(\frac{1}{k^{\gamma}}\right)$, $1 \le \gamma < 2$. The SPSA algorithms have appeared to be more efficient for small n than the SFDA approach. However, when the dimensionality of the task increases, the SFDA method becomes more efficient than SPSA algorithm according to the number of function value computed for optimization.

The linear estimator for the minimum value of optimized function has been proposed, using the theory of order statistics, and studied in experimental way. The estimator proposed are simple and depend only on the parameter of the extreme value distribution α. The parameter α is easily estimated, using the parameter of homogeneity of the objective function. Theoretical considerations and computer examples have shown that the confidence interval of the function minimum can be estimated with an admissible accuracy, when the number of iterations is increased. Finally, the developed algorithms were applied to the minimization of the mean absolute pricing error for parameter estimation in the Heston stochastic volatility model and minimization of weight of bulkheads for cargo oil tanks demonstrate applicability for practical purposes.

References

1. Yu. M. Ermoliev, *Methods of stochastic programming*, Nauka, Moscow (in Russian) (1976).
2. O. N. Granichin and B. T. Poliak, *Randomized algorithms for estimation and optimization with almost arbitrary errors*, Nauka, Moskow (in Russian) (2003).
3. H. J. Kushner and G. G. Yin, *Stochastic Approximation and Recursive Algorithms and Applications*, Springer, N.Y., Heidelberg, Berlin (2003).
4. V. S. Mikhalevitch, A. M. Gupal, V. I. Norkin, *Methods of Nonconvex Optimization*, Nauka, Moscow (in Russian) (1987).
5. J. C. Spall, *Introduction to Stochastic Search and Optimization: Estimation, Simulation, and Control*, Wiley, Hoboken, NJ (2003).
6. M. T. Vazan, Stochastic approximation, *Transactions in Mathematics and Mathematical Physics,* Cambridge University Press, Cambridge (1969).
7. F. H. Clarke, Generalized gradients and applications, *Trans. Amer. Math. Soc* **205**, 247-262 (1975).
8. V. Bartkutė and L. Sakalauskas, Convergence of simultaneous perturbation stochastic approximation in a Lipshitz class, *Lith. Mathematical Journal* **44**, 603-608, (in Lithuanian) (2004).
9. J. Mockus, *Multiextremal problems in design*, Nauka, Moscow (1967).
10. A. Žilinskas, A. Zhigljavsky, *Methods of the Global Extreme Searching*, Nauka, Moscow (In Russian) (1991).
11. A. Zhigljavsky, Branch and probability bound methods for global optimization. *Informatica* **1**(1), 125-140 (1990).
12. H. Chen, Estimation of the location of the maximum of a regression function using extreme order statistics, *Journal of multivariate analysis* **57**, 191-214 (1996).
13. P. Hall, On estimating the endpoint of a distribution, *Annals of Statistics* **10**, 556-568 (1982).
14. S. L. Heston, A closed-form solution for options with stochastic volatility with applications to bond and currency options, *The Review of Financial Studies* **6** (2), 327-343 (1993).
15. G. V. Reklaitis, A. Ravindran, K. M. Ragsdell, *Engineering Optimization. Methods and Applications*, Moscow. (In Russian) (1986).

APPLICATION OF THE MONTE-CARLO METHOD TO STOCHASTIC LINEAR PROGRAMMING

L. SAKALAUSKAS AND K. ŽILINSKAS

Institute of Mathematics and Informatics, Akademijos st 4
08663 Vilnius, Lithuania, E-mail: sakal@ktl.mii.lt, kest.zil@gmail.com

In this paper the method by a finite sequence of Monte-Carlo sampling estimators has been developed to solve stochastic linear problems. The method is grounded by adaptive regulation of the size of Monte-Carlo samples and the statistical termination procedure, taking into consideration the statistical modeling error. Our approach distinguishes itself by treatment of the accuracy of the solution in a statistical manner, testing the hypothesis of optimality according to statistical criteria, and estimating confidence intervals of the objective and constraint functions. To avoid "jamming" or "zigzagging" solving the problem, we implement the ε–feasible direction approach. The adjustment of sample size, when it is taken inversely proportional to the square of the norm of the Monte-Carlo estimate of the gradient, guarantees the convergence a. s. at a linear rate. The numerical study and examples in practice corroborate the theoretical conclusions and show that the procedures developed make it possible to solve stochastic problems with a sufficient agreeable accuracy by means of the acceptable amount of computations.

1. Introduction

Stochastic programming deals with a class of optimization models in which some of the data may be subject to significant uncertainty. Such models are appropriate when data evolve over time and decisions have to be made prior to observing the entire data streams. Although widespread applicability of stochastic programming models has attracted considerable attention of researchers, stochastic linear models remain one of the more challenging optimisation problems. Methods based on the approximation and decomposition are often applied to solve stochastic programming tasks,[1,2,3] however such ones can lead to very large-scale problems, and, thus, require very large computational resources. The study of stochastic programming algorithms therefore led to alternative ways of approximating problems, some of which obey certain asymptotic properties. This reliance on approximations has prompted to study the asymptotic convergence of solutions of approximate problems to a solution of original,[4,2] and consider adaptive methods for approximations.[5,6] In this paper we develop an adaptive approach for solving the stochastic linear problems by the Monte-Carlo method, based on asymptotic properties of Monte-Carlo sampling estimators. An approach is grounded by the treatment of statistical simulation error in a statistical manner and the rule for iterative regulation of the size of Monte-Carlo samples.

We consider a two-stage stochastic optimization problem with complete recourse:

$$F(x) = c \cdot x + E\{Q(x, w)\} \to \min_{x \in D} \tag{1}$$

subject to the feasible set

$$D = \left\{ x \middle| \; A \cdot x = b, \; x \in \Re_{+}^{n} \right\}, \tag{2}$$

where

$$Q(x, \omega) = \min{}_y [q \cdot y \,|\, W \cdot y + T \cdot x \le h, y \in R_{+}^{m}], \tag{3}$$

the vectors b, q, h and matrices A, W, T are of the appropriate dimensionality. Assume vectors q, h and matrices W, T random in general, and, consequently, depending on an elementary event $\omega \in \Omega$ from certain probability space (Ω, Σ, P). Thus, under uncertainty the modelled system operates in an environment in which there are uncontrollable parameters, which are modelled using random variables. Hence, the performance of such a system can also be viewed as a random variable. Let the measure P be absolutely continuous and defined by the probability density function $p(\cdot)$, i.e., the randomness is supposed to be exogenous, that cannot be affected by decision. Besides, assume that a solution of the second stage problem (3) and values of function Q almost surely (a.s.) exist and are bounded.

Stochastic procedures for solving problems of this kind are often considered and two ways are used to achieve the convergence of developed methods. The first one leads to the class of methods of stochastic approximation. The convergence in stochastic approximation is ensured by regulating certain step-length multipliers in a scheme of stochastic gradient search.[7,8] The convergence of these methods can be improved in case more precise estimators used on semi-stochastic approximation.[9,10] However, the methods of stochastic approximation converge rather slowly and, besides, it is not so clear how to terminate the process of stochastic approximation.

The second way to ensure the convergence in stochastic optimisation is related to application of the methods of a relative stochastic gradient error. The theoretical scheme of such methods requires that the variance of the stochastic gradient be varied in the optimisation procedure so that to remain proportional to the square of the gradient norm.[11] This approach offers an opportunity to develop implementable algorithms of stochastic optimisation.[12,13] We consider them here using a finite series of Monte-Carlo estimators for algorithm construction in stochastic linear programming.

2. Stochastic Differentiation and Monte-Carlo Estimators

The gradient search is the most often used way of constructing methods for numerical optimisation. Since mathematical expectations in (1) are computed explicitly only in rare cases, it is complicated all the more to analytically compute the gradients of functions, containing this expression. The Monte-Carlo method is a universal and convenient tool of estimating these expectations and we try to apply it to estimate derivatives, too. The procedures of gradient evaluation are often constructed by expressing a gradient as an expectation and then evaluating this expectation by means of statistical simulation.[4,14,15,12]

First, by duality of linear programming we have that

$$F(x) = c \cdot x + E\left\{\max_u \left[(h - T \cdot x) \cdot u \mid u \cdot W^T + q \ge 0,\ u \in \Re^n_+\right]\right\}. \quad (4)$$

It can be derived, that under the assumption on the existence of a solution to the second stage problem in (3) and continuity of measure P, the objective function (4) is smoothly differentiable and its gradient is expressed as

$$\nabla_x F(x) = E(g(x,\omega)), \quad (5)$$

where $g(x,\omega) = c - T \cdot u^*$ is given by the a set of solutions of the dual problem

$$(h - T \cdot x)^T \cdot u^* = \max_u [(h - T \cdot x)^T \cdot u \mid u \cdot W^T + q \ge 0, \quad u \in \Re^m],$$

(details are given in[4,15]).

In solving problem (1), suppose it is possible to get finite sequences of realizations (trials) of ω at any point x and the corresponding solutions of problem (3), and the values of $Q(x,\omega)$ as well as solution of the second stage problem in (3) are available for these realizations. Then it is not difficult to find the Monte-Carlo estimators corresponding to the expectations in (1), (4), (5). Thus, we assume here that the Monte-Carlo samples of a certain size N are provided for any $x \in R^n$

$$Y = (y^1, y^2, \ldots, y^N), \tag{6}$$

where y^i are independent random variables, identically distributed with the density $p(\cdot) : \Omega \to R_+^n$, and the sampling estimators are computed:

$$\tilde{F}(x) = \frac{1}{N}\sum_{j=1}^{N} f(x, y^j) \quad , \tag{7}$$

$$\tilde{D}^2(x) = \frac{1}{N-1}\sum\nolimits_{i=1}^{N} (f(x, y^j) - \tilde{F}(x))^2 \quad . \tag{8}$$

The estimate of a gradient:

$$\tilde{G}(x) = \frac{1}{N}\sum\nolimits_{j=1}^{N} g(x, y^j) \quad , \tag{9}$$

and the sampling covariance matrix

$$Z(x) = \frac{1}{N-n}\sum\nolimits_{j=1}^{N} (g(x, y^j) - \tilde{G}) \cdot (g(x, y^j) - \tilde{G})' \tag{10}$$

will be of use later on.

3. Stochastic Procedure for Optimisation

Since in the stochastic optimization only the first order methods are working, we have confined ourselves by the gradient-descent type methods and show that typical deterministic approaches of constrained optimization might be generalized to the stochastic case. To avoid problems of "jamming" or "zigzagging" appearing in gradient search we implement the ε–feasible direction approach.

Let us define the set of *feasible directions* as follows:

$$V(x) = \left\{ g \in \Re^n \middle| Ag = 0, \forall_{1 \le i \le n} \left(g_j \le 0, \; if \; x_j = 0 \right) \right\}, \tag{11}$$

where g_U is assumed as projection of vector g onto the set U. Since the objective function is differentiable, the solution $x \in D$ is optimal if

$$\nabla F(x)_V = 0 \,. \tag{12}$$

Assume a certain multiplier $\hat{\rho} > 0$ to be given. Define the function $\rho_x : V(x) \to \Re_+$ by

$$\rho_x(g) = \min\left\{ \hat{\rho},\ \min_{\substack{g_j>0,\\ 1\le j\le n}}\left(\frac{x_j}{g_j}\right) \right\},\ \exists_{1\le j\le n}\left(g_j > 0\right), \tag{13}$$

$\rho_x(g) = \hat{\rho}$, if $\forall_{1\le j\le n}(g_j \le 0)$. Thus, $x + \rho \cdot g,\ x \in X$, when $\rho = \rho_x(g)$, for any $g \in V$. Now, let a certain small value $\hat{\varepsilon} > 0$ be given. Then we introduce the function $\varepsilon_x : V(x) \to \Re_+$

$$\varepsilon_x(g) = \hat{\varepsilon} \cdot \max_{\substack{1\le j\le n\\ g_j>0}}\left\{\min\left\{x_j, \hat{\rho}\cdot g_j\right\}\right\},\ \exists_{1\le j\le n}\left(g_j > 0\right),$$

$\varepsilon_x(g) = 0$, if $\forall_{1\le j\le n}(g_j \le 0)$, and define the ε *-feasible set*

$$V_\varepsilon(x) = \left\{ g \in \Re^n \middle| Ag = 0,\ \forall_{1\le i\le n}\left(g_j \le 0,\ if\ \left(0 \le x_j \le \varepsilon_x(g)\right)\right)\right\} \tag{14}$$

Let us start developing the stochastic optimization procedure. Assume certain initial point $x^0 \in D$ be given, random sample (6) of a certain initial size N^0 be generated at this point, and Monte-Carlo estimates (7), (8), (9), (10) be computed. For instance, the starting point can be obtained as the solution of the deterministic linear problem:

$$\begin{aligned}(x^0, y^0) = \arg\min_{x,y}[c\cdot x + q\cdot y \mid A\cdot x = b,\ W\cdot y + T\cdot x \le h,\\ y \in R_+^m, x \in R_+^n].\end{aligned} \tag{15}$$

The iterative stochastic procedure of gradient search could be used further:

$$x^{t+1} = x^t - \rho^t \cdot \tilde{G}(x^t). \tag{16}$$

where $\tilde{G}^t = \tilde{G}(x^t)$, $\rho^t = \rho_{x^t}(\tilde{G}^t)$ is a certain step-length multiplier defined by (13).

Let us consider the choice of the Monte-Carlo sample size more in detail. Note, that there is no great necessity to compute estimators with a high accuracy on starting the optimisation, because then it suffices only to approximately evaluate the direction leading to the optimum. Therefore, one can obtain not so large samples at the beginning of the optimum search and, later on, increase the size of samples so as to get the estimate of the objective function with a desired accuracy just at the time of decision making on finding the solution to the optimisation problem. We can pursue this purpose by choosing the sample size at every next iteration inversely proportional to the square of the gradient estimator from the current iteration:

$$N^{t+1} \geq \frac{\hat{\rho} \cdot C}{\rho^t \cdot \left|\tilde{G}^t\right|^2}, \tag{17}$$

where $C > 0$ is a certain constant, $\rho^t = \rho_{w^t}(\hat{G}^t)$, $\tilde{G}^t$ is an ε-feasible direction at the point x^t (i.e., the projection of gradient estimate (9) to the ε-feasible set (14)). On the other hand, such a rule enables us to ensure the condition of proportionality of stochastic gradient variance to the square of the gradient norm, which is sufficient for convergence.[11] Thus, under certain wide conditions on existence of expectations of estimators such the rule guarantees the convergence a.s. to optimal solution with linear rate, i.e., starting from any initial approximation $x^0 \in D$ and $N^0 > 1$, formulae (15), (16), (17) define the sequence $\left\{x^t, N^t\right\}_0^\infty$ so that $x^t \in D$, and there exist values $\overline{\rho} > 0$, $\varepsilon_0 > 0$, $\overline{C} > 0$ such that

$$\lim_{t\to\infty}\left|\nabla F(x^t)_{V_\varepsilon^t}\right|^2 = 0 \quad (\mathrm{mod(P)}), \tag{18}$$

For $0 < \hat{\rho} \leq \overline{\rho}$, $0 < \varepsilon < 1$, $C \geq \overline{C}$. The proof is available.[13]

Let us discuss a choice of parameters of the method. The step length ρ in (16) can be determined experimentally. The choice of constant C or that of the best metrics for computing the stochastic gradient norm in (16) requires a separate study. For instance, the choice $C = n \cdot Fish(\gamma, n, N^t - n) \approx \chi_\gamma^2(n)$, where $Fish(\gamma, n, N^t - n)$ is the γ-quantile of the Fisher distribution with $(n, N^t - n)$ degrees of freedom, and estimation of the gradient norm in a metric induced by the sampling covariance matrix (10), ensure that a random error of the stochastic gradient does not exceed the gradient norm approximately with probability $1 - \gamma$. Thus, we propose a following version of (17) for regulating the sample size in practice:

$$N^{t+1} = \min\left(\max\left(\left[\frac{n \cdot Fish(\gamma, n, N^t - n)}{\rho^t \cdot (\tilde{G}(x^t) \cdot (Z(x^t))^{-1} \cdot (\tilde{G}(x^t)}\right] + n, N_{\min}\right), N_{\max}\right). \tag{19}$$

Minimal N_{min} (usually ~20-50) and maximal N_{max} (usually ~ 1000-2000) values are introduced to avoid great fluctuations of sample size in iterations. Note that N_{max} also may be chosen from the conditions on the permissible confidence interval of estimates of the objective function.

4. Statistical Testing of the Optimality Hypothesis

A possible decision on finding of optimal solution should be examined at each step of the optimization process. Since we know only the Monte-Carlo estimates of the objective function and that of its gradient, we can test only the statistical optimality hypothesis. As far as the stochastic error of these estimates depends in essence on the Monte-Carlo samples size, a possible optimal decision could be made, if, first, there is no reason to reject the hypothesis of equality to zero of the gradient, and, second, the sample size is sufficient to estimate the objective function with the desired accuracy.

Note that the distribution of sampling averages (7) and (9) can be approximated by the one- and multidimensional Gaussian laws. Therefore it is convenient to test the validity of the stationarity condition (12) by means of the well-known multidimensional Hotelling T^2-statistics. Hence, the optimality hypothesis could be accepted for some point x^t with significance $1-\mu$, if the following condition is satisfied:

$$(N^t - n) \cdot (\tilde{G}(x^t)) \cdot (Z(x^t))^{-1} \cdot (\tilde{G}(x^t)) / n \le Fish(\mu, n, N^t - n) \quad . \tag{20}$$

Next, we can use the asymptotic normality again and decide that the objective function is estimated with a permissible accuracy ε, if its confidence bound does not exceed this value:

$$2 \cdot \eta_\beta \cdot \tilde{D}(x^t) / \sqrt{N^t} \le \varepsilon \quad , \tag{21}$$

where η_β is the β -quantile of the standard normal distribution. Thus, the procedure (11) is iterated adjusting the sample size according to (19) and testing conditions (20) and (21) at each iteration. If the latter conditions are met at some iteration, then there are no reasons to reject the hypothesis on the optimum finding. Therefore, there is a basis to terminate the optimization and make a decision on the optimum finding with a permissible accuracy. If at least one condition out of (20), (21) is unsatisfied, then the next sample is generated and the optimization is continued. As it follows from the previous section, the optimization should terminate after generating a finite number of Monte-Carlo samples.

5. Computer Study

Let us consider an application of the approach developed to example taken from literature. In this example we consider the two-stage stochastic linear optimisation problem. Data of the problem are taken from a database at the address http://www.math.bme.hu/~deak/twostage/l1/20x20.2.

Dimensions of the task are as follows: the first stage has 10 rows and 20 variables; the second stage has 20 rows and 30 variables.

The estimate of the optimal value of the objective function given in the database is 182.94234 ± 0.066. Application of the approach considered allows us to improve the estimate of the optimal value up to 182.59248 ± 0.03300.

Now, let us consider the results, obtained in solving this task 400 times by the method (15), (16), (19). Initial data were as follows: $\gamma = \beta = 0.95$, $\mu = 0.99$, $\varepsilon = 2$, $N^0 = N_{min} = 100$, maximal number of iterations $t_{\max} = 50$, generation of trials was broken when the estimated confidence interval of the objective function exceeds admissible value ε.

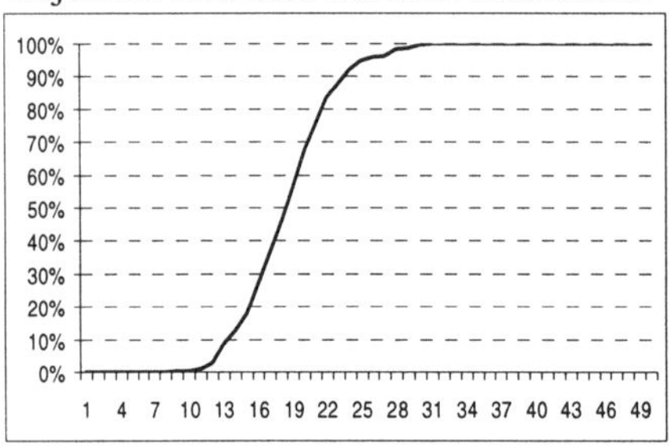

Figure 1. Frequency of stopping.

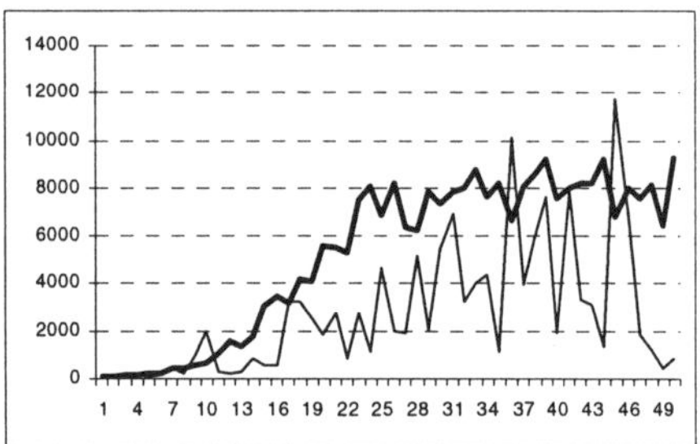

Figure 2. Change of the sample size.

Termination conditions a) - c) were satisfied at least once time for all paths of optimization. Thus, the conclusion on the optimum finding with an admissible accuracy could be made for all paths (the sampling frequency of termination after t iterations with confidence intervals is presented in Fig. 1). The averaged dependencies of the sample size, the objective function, the confidence interval of (21), the Hotelling statistic of (20) by the iteration number t are given in Fig's.

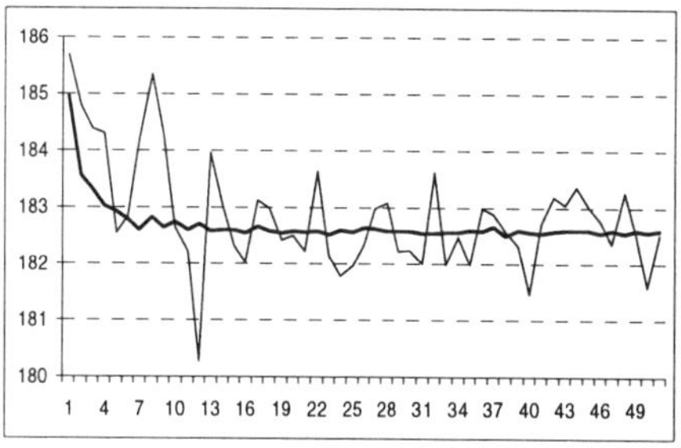

Figure 3. Change of the objective function.

— path ▬ averaged

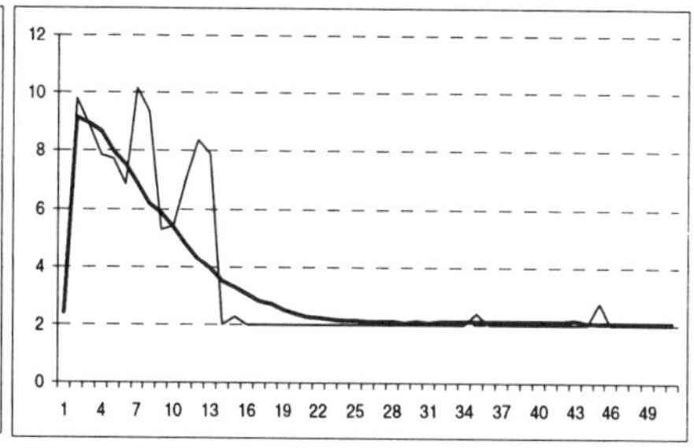

Figure 4. Change of confidence interval.

— path ▬ averaged

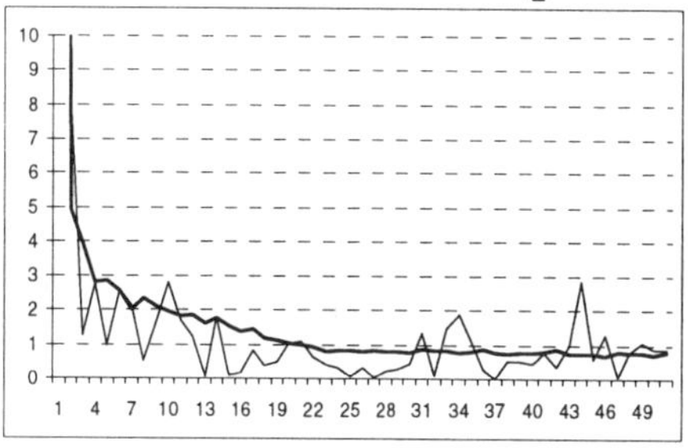

Figure 5. Change of the Hotelling statistics.

— path ▬ averaged

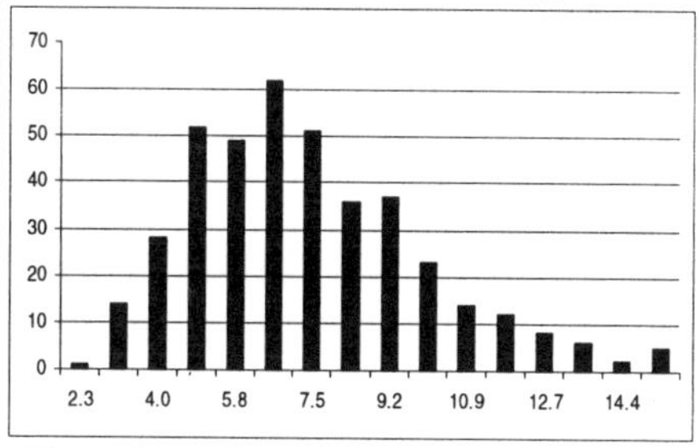

Figure 6. Histogram of ratio $\sum_{j=1}^{t} N^j / N^t$

2-5 to illustrate the convergence and the behavior of the optimization process. Also, one path of realization of the optimization process illustrates the stochastic character of this process in these figures. In Fig.6 is a histogram of the ratio $\sum_{j=1}^{t} N^j / N^t$ is depicted.

6. Discussion and Conclusions

Thus, the stochastic iterative method has been developed to solve the stochastic linear programming problems by a finite sequence of Monte-Carlo sampling estimators. The approach presented is grounded by the stopping procedure and the rule for adaptive regulation of size of Monte-Carlo samples, taking into account the statistical modeling error. The termination procedure proposed allows us to test the optimality hypothesis and to evaluate the confidence intervals of the objective and constraint functions in a statistical way. The regulation of sample size inversely proportional to the square of the norm of gradient of the Monte-Carlo estimator allows us to solve SLP problems rationally from the computational viewpoint and guarantees the convergence a.s. at a linear rate. The numerical study and an example in practice corroborate the theoretical conclusions and show that the procedures developed make it possible to solve stochastic problems with a sufficient agreeable accuracy by means of the acceptable amount of computations.

References

1. Yu. Ermolyev, *Methods of Stochastic Programming*, Nauka, Moscow, (in Russian) (1976).
2. A. Prekopa, *Stochastic Programming*, Kluwer (1995).
3. K. Marti, *Stochastic Optimization Methods*, Springer, N.Y. (2005).
4. R. Rubinstein and A. Shapiro, *Discrete Events Systems: Sensitivity Analysis and Stochastic Optimization by the Score Function Method*, Wiley & Sons, N.Y. (1993).
5. A. Shapiro and T. Homem-de-Mello, A simulation-based approach to two-stage stochastic programming with recourse, *Mathematical Programming* **81**, 301-325 (1998).
6. J. L. Higle and S. Sen, Statistical approximations for stochastic linear programming problems, *Annuals of Operations Research* **85**, 173-192 (1999).
7. V. S. Mikhalevitch, A. M. Gupal and V. I. Norkin, *Methods of Nonconvex Optimization*, Nauka, Moscow (in Russian) (1987).
8. H. J. Kushner and G. G. Yin, *Stochastic Approximation and Recursive Algorithms and Applications*, Springer, N.Y., Heidelberg, Berlin (2003).
9. K. Marti, Descent stochastic quasigradient methods, In: Yu. Ermolyev and R. Wets (eds.), *Numerical Techniques for Stochastic Optimization*, Springer-Verlag, Berlin, pp. 393-400 (1988).
10. K. Marti, Optimal semi-stochastic approximation procedures. *ii. Z. Angew. Math. Mech.* **69**, 67-69 (1989).
11. B. T. Polyak, *Introduction to Optimization*, Translations Series in Mathematics and Engineering, Optimization Software, Inc., Publications Division, New York (1987).
12. L. Sakalauskas, Nonlinear stochastic programming by Monte-Carlo estimators. *European Journal on Operational Research* **137**, 558-573 (2002).
13. L. Sakalauskas, Application of the Monte-Carlo method to nonlinear stochastic optimization with linear constraints, *Informatica* **15**(2), 271-282 (2004).
14. K. Marti, Differentiation formulas for probability functions: the transformation method, *Mathematical Programming*, **75**(2), 201-220 (1996).
15. A. Shapiro, Stochastic Programming by Monte-Carlo simulation methods. *Stochastic Programming E-Print Series* (2000).

STUDYING THE RATE OF CONVERGENCE OF THE STEEPEST DESCENT OPTIMISATION ALGORITHM WITH RELAXATION

R. J. HAYCROFT

Cardiff University, School of Mathematics, Senghennydd Road Cardiff CF24 4AG, UK, E-mail: HaycroftRJ@cf.ac.uk

Gradient-type algorithms can be linked to algorithms for constructing optimal experimental designs for linear regression models. The asymptotic rate of convergence of these algorithms can be expressed through the asymptotic behaviour of an experimental design construction procedure. One well known gradient-type algorithm is the method of Steepest Descent. Here a generalised version of the Steepest Descent algorithm, with a relaxation coefficient is considered and the rate of convergence of this algorithm is investigated.

keywords: Gradient optimisation; Steepest Descent

1. Introduction

Consider the problem of minimising an objective function $f(x)$ in $\mathbb{R}^d$. Most numerical minimisation methods take an iterative form. An initial point $x^{(0)}$ is first selected, from which a sequence of points $x^{(k)}$, $k = 1, 2, \ldots$ can then be produced. As the sequence progresses, each new point represents an improved approximation of the minimum point.

Iterative algorithms can be written in the following general form

$$x^{(k+1)} = x^{(k)} - s^{(k)} d^{(k)} , \tag{1}$$

where $d^{(k)}$ is the direction down which the next point is to be selected and $||s^{(k)} d^{(k)}||$ is the step size taken in the chosen direction.

Gradient methods, the most famous of which being the method of Steepest Descent, are one such family of optimisation algorithms. Here the direction $d^{(k)}$ in (1) is the gradient of the objective function. Thus for a general smooth function $f(x)$ in $\mathbb{R}^d$ gradient algorithms can be written as

$$x^{(k+1)} = x^{(k)} - \alpha^{(k)} \bigtriangledown f(x^{(k)}) , \tag{2}$$

where $\nabla f(x^{(k)}) = g(x^{(k)}) = \left(\frac{\partial f}{\partial x_1}, \ldots, \frac{\partial f}{\partial x_d}\right)^T$ and $\alpha^{(k)}$ is the step length. Gradient algorithms of the form (2) are classical in the theory and practise of optimisation, see for example, Chapter 7 of [1].

Convergence rates of gradient-type algorithms are limited by the so called Kantorovich bounds, see [2]. The asymptotic behaviour of the Steepest Descent algorithm has been thoroughly studied in Chapter 7 of [3] and extended in [4] and [5]. The first asymptotic results concerning the asymptotic behaviour of the steepest descent algorithm with relaxation have been described in Sec. 7.4 of [3], but that study was only done for $d = 2$ and was mainly constrained to simulation.

Gradient algorithms for quadratic functions are introduced in Sec. 2. A general renormalised updating formula for gradient algorithms is also given along with an explanation of the renormalisation process involved. The steepest descent algorithm with relaxation coefficient, together with a proof of the convergence of the algorithm to a probability distribution, is then presented in Sec. 3. A study was undertaken into the dependence of the asymptotic rate of convergence on the choice of relaxation coefficient γ, and on the eigenvalues involved in the problem. The results of this study, including some graphical representations are shown in Sec. 4.

2. The General Quadratic Case

Consider the problem of finding the minimum value x^* of a quadratic function

$$f(x) = \frac{1}{2}x^T Ax - x^T b$$

where A is a positive definite symmetric matrix with eigenvalues

$$0 < m = \lambda_1 \leq \lambda_2 \leq \ldots \leq \lambda_d = M < \infty\,.$$

Without loss of generality assume that

$$A = \begin{pmatrix} \lambda_1 & 0 & \ldots & 0 \\ 0 & \lambda_2 & & 0 \\ \vdots & & \ddots & \vdots \\ 0 & 0 & \ldots & \lambda_d \end{pmatrix}, \qquad (3)$$

since a change of variables can always achieve this form. The gradient of $f(x)$ is

$$g(x) = Ax - b\,.$$

We can also assume that $x^* = 0$ (otherwise the substitution $x - x^* \to x$ can be made). Gradient algorithms then take the following general form

$$x^{(k+1)} = x^{(k)} - \alpha^{(k)} g^{(k)} \tag{4}$$

where $\alpha^{(k)}$ is the step length at iteration k and $g^{(k)} = Ax^{(k)} - b$. By multiplying both sides of the equation by A the following form of the equation can be obtained

$$g^{(k+1)} = g^{(k)} - \alpha^{(k)} A g^{(k)} . \tag{5}$$

By re-normalising the gradient $g^{(k)}$, according to

$$z^{(k)} = \frac{g^{(k)}}{||g^{(k)}||} = \frac{g^{(k)}}{(g^{(k)}, g^{(k)})^{1/2}}$$

(so that $||z^{(k)}|| = 1$ for all $k = 1, 2, ...$), algorithms of this type can be related to dynamical systems. Once in this form, information about the asymptotic rates of convergence of the algorithms can be more readily obtainable.

Since matrix A has the form (3), (5) can be re-written component-wise as

$$g_i^{(k+1)} = g_i^{(k)} - \alpha^{(k)} \lambda_i g_i^{(k)} \quad \text{for } i = 1, \ldots, d.$$

Denote the i-th component of $z^{(k)}$ as w_i and the i-th component of $z^{(k+1)}$ as $w_i^{'}$:

$$z^{(k)} = (w_1, ..., w_d)^T, \qquad z^{(k+1)} = (w_1^{'}, ..., w_d^{'})^T;$$

set also $\alpha = \alpha^{(k)}$, $p_i = w_i^2$, $p_i^{'} = (w_i^{'})^2$. Then

$$p_i = \frac{(g^{(k)})_i^2}{\sum_j (g^{(k)})_j^2}, \qquad p_i^{'} = \frac{(g^{(k+1)})_i^2}{\sum_j (g^{(k+1)})_j^2} \tag{6}$$

where $\sum p_i = 1$, $\sum p_i^{'} = 1$, $p_i, p_i^{'} \geq 0$. The algorithm (5) then implies the following updating formula for p_i :

$$p_i^{'} = \frac{(1 - \alpha\lambda_i)^2}{1 - 2\alpha\mu_1 + \alpha^2\mu_2} p_i , \tag{7}$$

We can consider (7) as an updating formula:

$$P^{'} = \Phi(P) \tag{8}$$

for the probability distribution P with P and $P^{'}$ supported on $\lambda_1, ..., \lambda_d$ with weights $p_1, ..., p_d$ and $p_1^{'}, ..., p_d^{'}$ respectively. Since in (7) $P = P^{(k)}$ and

$P' = P^{(k+1)}$ (the distributions at iteration k and $(k+1)$) we can also write (8) as

$$P^{(k+1)} = \Phi(P^{(k)}) . \tag{9}$$

The moments of these distributions are given by

$$\mu_l = \mu_l(P) = \sum_{i=1}^{d} \lambda_i^l p_i , \qquad \mu_l' = \mu_l(P') = \sum_{i=1}^{d} \lambda_i^l p_i' .$$

Let us express the first two moments of P' (that is, μ_1' and μ_2') through the moments of P.

First let us check that $\mu_0' = \sum p_i' = 1$. Indeed,

$$\mu_0' = \sum_{i=1}^{d} p_i' = \frac{1}{1 - 2\alpha\mu_1 + \alpha^2\mu_2} \sum_{i=1}^{d} (1 - \alpha\lambda_i)^2 p_i$$
$$= \frac{1}{1 - 2\alpha\mu_1 + \alpha^2\mu_2} \left[1 - 2\alpha\mu_1 + \alpha^2\mu_2 \right] = 1 .$$

μ_1' and μ_2' expressed through the moments of P are then

$$\mu_1' = \sum_{i=1}^{d} \lambda_i p_i' = \frac{1}{1 - 2\alpha\mu_1 + \alpha^2\mu_2} \left[\mu_1 - 2\alpha\mu_2 + \alpha^2\mu_3 \right] ,$$

$$\mu_2' = \sum_{i=1}^{d} \lambda_i^2 p_i' = \frac{1}{1 - 2\alpha\mu_1 + \alpha^2\mu_2} \left[\mu_2 - 2\alpha\mu_3 + \alpha^2\mu_4 \right] .$$

A rate of convergence at iteration k in this family of algorithms is defined by

$$r^{(k)} = \frac{(g^{(k+1)}, g^{(k+1)})}{(g^{(k)}, g^{(k)})} = \frac{(g^{(k)}, g^{(k)}) - 2\alpha(Ag^{(k)}, g^{(k)}) + \alpha^2(A^2 g^{(k)}, g^{(k)})}{(g^{(k)}, g^{(k)})}$$
$$= 1 - 2\alpha\mu_1 + \alpha^2\mu_2 . \tag{10}$$

This rate corresponds to the denominator of the updating formula (7).

3. The Steepest Descent Algorithm

The simplest of the gradient methods is the method of Steepest Descent. The step length $\alpha^{(k)}$ is chosen so that, at iteration $(k+1)$, the function takes on the minimum value possible along direction $\nabla f(x^{(k)})$. For a general convex function the steepest descent algorithm is

$$x^{(k+1)} = x^{(k)} - \alpha^{(k)} \nabla f(x^{(k)}) ,$$

where

$$\alpha^{(k)} = \arg\min_{\alpha} f(x^{(k)} - \alpha\nabla f(x^{(k)})) \, .$$

This can be rewritten as

$$x^{(k+1)} = x^{(k)} - \frac{(g^{(k)}, g^{(k)})}{(Ag^{(k)}, g^{(k)})} g^{(k)} \, . \qquad (11)$$

The method of steepest descent can be considered desirable on account of its simplicity, easy application and relative stability as an algorithm. Unfortunately it generally has a poor convergence rate. There are however, adaptations to improve the performance of this algorithm. The addition of a relaxation coefficient γ, where $0 < \gamma < \frac{2M}{m+M}$, to the steepest descent algorithm, i.e.

$$x^{(k+1)} = x^{(k)} - \gamma\frac{(g^{(k)}, g^{(k)})}{(Ag^{(k)}, g^{(k)})} g^{(k)} \qquad \text{with } \alpha^{(k)} = \gamma\frac{(g^{(k)}, g^{(k)})}{(Ag^{(k)}, g^{(k)})} \, , \qquad (12)$$

will radically alter the behaviour of the algorithm. The renormalised algorithm without relaxation coefficient γ converges to a two-point cycle but with $\gamma \neq 1$ the algorithm may possess a much better asymptotic convergence rate. For some γ close to 1 the algorithm can exhibit chaotic behaviour.

As $\alpha^{(k)} = \frac{\gamma}{\mu_1}$, we obtain

$$p_i^{'} = \frac{(1-\alpha\lambda_i)^2}{1-2\alpha\mu_1+\alpha^2\mu_2} p_i = \frac{(1-\frac{\gamma}{\mu_1}\lambda_i)^2}{1-2\gamma+\gamma^2\frac{\mu_2}{\mu_1^2}} p_i \, , \qquad (13)$$

$$\mu_1^{'} = \frac{\mu_1^3 - 2\gamma\mu_1\mu_2 + \gamma^2\mu_3}{\gamma^2\mu_2 + \mu_1^2(1-2\gamma)} \quad \text{and} \quad \mu_2^{'} = \frac{\mu_1^2\mu_2 - 2\gamma\mu_1\mu_3 + \gamma^2\mu_4}{\gamma^2\mu_2 + \mu_1^2(1-2\gamma)} \, .$$

Theorem. *Let $P^{(0)}$ be any non-degenerate probability measure with support $\{\lambda_1, ..., \lambda_d\}$ and let the sequence of probability measures $\{P^{(k)}\}$ be defined via the updating formula (8). Then the sequence*

$$L^{(k)} = L(P^{(k)}) = \gamma\mu_2(P^{(k)}) - \mu_1^2(P^{(k)})$$

monotonously increases and converges for $\gamma > 1$ and $0 < \gamma < \frac{\mu_1^2}{\mu_2}$; that is $L^{(0)} \leq L^{(1)} \leq \ldots \leq L^{(k)} \leq \ldots$ and the limit $L^*(P^{(0)}) = \lim_{k\to\infty} L^{(k)}$ exists. If $P^{(0)}$ is such that $P^{(0)}(\lambda_1) > 0$ and $P^{(0)}(\lambda_d) > 0$, then the limit $\lim_{k\to\infty} L^{(k)}$ does not depend on $P^{(0)}$ and

$$L^*(P^{(0)}) = L^* = \frac{\gamma^2}{4}(m+M)^2 - mM \, ;$$

moreover, the sequence of probability measures $\{P^{(k)}\}$ *converges (as* $k \to \infty$*) to the probability measure* P^* *supported at the points* m *and* M *with weights*

$$P^*(m) = \frac{\gamma m - 2M + \gamma M}{2(m-M)} \quad \text{and} \quad P^*(M) = \frac{2m - \gamma m - \gamma M}{2(m-M)},$$

where $1 \leq \gamma \leq \frac{2M}{m+M}$. If $\gamma > \frac{2M}{m+M}$ then the optimisation algorithm (12) diverges.

Proof. $L^{(k)}$ is not positive if $0 < \gamma < \mu_1/\mu_2$. Note that $L^{(k)} \geq 0$ for any P in view of the Cauchy-Schwartz inequality.

The sequence $L^{(k)}$ is non-decreasing if $L^{(k+1)} - L^{(k)} \geq 0$ holds for any distribution P. If the distribution P is degenerate (that is, has mass 1 at one point) then $L^{(k+1)} = L^{(k)}$ and the statement of the theorem holds. We assume below that P is non-degenerate. In particular this implies $\mu_2 > \mu_1^2$. We have:

$$L^{(k+1)} - L^{(k)} \geq 0 \iff \left(\gamma\mu_2^{'} - (\mu_1^{'})^2\right) - \left(\gamma\mu_2 - \mu_1^2\right) \geq 0\,. \tag{14}$$

We can represent the left hand side of (14) as

$$\left(\gamma\mu_2^{'} - (\mu_1^{'})^2\right) - \left(\gamma\mu_2 - \mu_1^2\right) = \gamma\frac{U}{W}$$

with,

$$\begin{aligned} U = &-{\mu_1}^2{\mu_2}^2\gamma^2 + 4\gamma^2\mu_1\mu_3\mu_2 - 2\gamma^3\mu_1\mu_3\mu_2 - 4{\mu_1}^2{\mu_2}^2\gamma + 4{\mu_1}^4\mu_2 - 4\gamma{\mu_1}^3\mu_3 \\ &+\gamma^2\mu_4{\mu_1}^2 + 4{\mu_1}^4\mu_2\gamma + 4\gamma^2{\mu_1}^3\mu_3 + \gamma^4\mu_4\mu_2 - 2\gamma^3\mu_4{\mu_1}^2 + 5{\mu_1}^2{\mu_2}^2\gamma^3 \\ &-8{\mu_1}^4\mu_2\gamma^2 - \gamma^3{\mu_3}^2 - 4{\mu_1}^6 + 4{\mu_1}^6\gamma - \gamma^4{\mu_2}^3\,, \end{aligned}$$

and

$$W = \left(-\gamma^2\mu_2 - {\mu_1}^2 + 2\,\gamma{\mu_1}^2\right)^2\,.$$

It is clear W will always remain positive thus the problem is reduced to determining whether or not the numerator, U, is non-negative. One way of proving that an expression is non-negative is to relate it to a variance of a random variable, as it is known that variances are always non-negative. Consider the variance

$$V = var(a\xi + b\xi^2)$$

where ξ is the random variable with distribution P and a, b are some parameters we shall choose.

$$\begin{aligned} V = var(a\xi + b\xi^2) &= E(a\xi + b\xi^2)^2 - [E(a\xi + b\xi^2)]^2 \\ &= a^2E(\xi^2) + 2abE(\xi^3) + b^2E(\xi^4) - ([aE(\xi) + bE(\xi^2)])^2 \\ &= a^2\mu_2 + 2ab\mu_3 + b^2\mu_4 - (a\mu_1 + b\mu_2)^2 \,. \end{aligned}$$

Subtract the variance V from U and consider this as a function of a, b:

$$\begin{aligned} F(a,b) = U - V = {} & 4\gamma^2\mu_1\mu_3\mu_2 - \gamma^3{\mu_3}^2 - 4{\mu_1}^6 + 4\,{\mu_1}^6\gamma 4\,{\mu_1}^2{\mu_2}^2\gamma + 4\,{\mu_1}^4\mu_2 \\ & -4\,\gamma{\mu_1}^3\mu_3 + \gamma^2\mu_4{\mu_1}^2 - {\mu_1}^2{\mu_2}^2\gamma^2 + 4\,{\mu_1}^4\mu_2\gamma + 4\,\gamma^2{\mu_1}^3\mu_3 \\ & +\gamma^4\mu_4\mu_2 - 2\,\gamma^3\mu_4{\mu_1}^2 + 5\,{\mu_1}^2{\mu_2}^2\gamma^3 - 8\,{\mu_1}^4\mu_2\gamma^2 - 2\,\gamma^3\mu_1\mu_3\mu_2 \\ & -\gamma^4{\mu_2}^3 - a^2\mu_2 - 2\,ab\mu_3 - b^2\mu_4 + a^2{\mu_1}^2 + 2\,a\mu_1 b\mu_2 + b^2{\mu_2}^2 \,. \end{aligned}$$

Let us attempt to select a, b so that $F(a, b) = 0$, which would mean that $U = V$. First, choose b to eliminate the μ_4 term:

$$b = b_0 = \gamma\sqrt{\gamma^2\mu_2 + \mu_1^2 - 2\gamma\mu_1^2} \,.$$

It is easy to see that $\gamma^2\mu_2 + \mu_1^2 - 2\gamma\mu_1^2 \geq 0$ since by adding and subtracting $\gamma^2\mu_1^2$ we obtain

$$\gamma^2\mu_2 + (\mu_1^2 - 2\gamma\mu_1^2 + \gamma^2\mu_1^2) - \gamma^2\mu_1^2 = \mu_1^2(1-\gamma)^2 + \gamma^2(\mu_2 - \mu_1^2)$$

which is obviously non-negative and so b is always correctly defined. Substitute this value into $F(a, b)$, then solve $F(a, b_0) = 0$ with respect to a. $F(a, b_0)$ is a quadratic function in a. Let D be the discriminant of $F(a, b_0)$. This discriminant can be simplified to

$$D = (\gamma - 1)\left(\gamma\mu_2 - {\mu_1}^2\right)\left(\mu_3\gamma + 2\,{\mu_1}^3 - \gamma\mu_1\mu_2 - 2\,\mu_2\mu_1\right)^2$$

This is clearly non-negative for $\gamma > 1$ and $0 < \gamma < \frac{\mu_1^2}{\mu_2}$ since $\mu_2 - \mu_1^2 \geq 0$ by the Cauchy-Schwartz inequality. There is therefore a solution of the equation $F(a, b) = 0$ and hence the conclusion can be drawn that $L^{(k)}$ monotonously increases and converges to a limit i.e. $\lim_{k\to\infty} L^{(k)}$ exists. □

When $\frac{\mu_1^2}{\mu_2} < \gamma < 1$ the relaxed steepest descent algorithm does not converge to an optimum design. It is within this range of γ that improved asymptotic rates of convergence are demonstrated. The asymptotic rate of convergence is defined to be

$$R = e^{1/N \sum ln(r^{(k)})} \,. \tag{15}$$

The worst rate, achieved when $L^{(k)}$ converges to L^*, is given by

$$R_{worst} = \left(\frac{M-m}{M+m}\right)^2 .$$

4. Results

Figure 1 is a typical depiction of R for the relaxed steepest descent algorithm as a function of γ. Here fastest convergence occurs when γ is approximately 0.95 but this varies depending on the eigenvalues.

In Fig. 2 the asymptotic rates of convergence for three values of γ are plotted as functions of M. The general trend shows as M increases the rate worsens but not nearly as fast as R_{worst} worsens. Middle eigenvalues were chosen so that even spacing between eigenvalues is accomplished. Numerical results show that this situation is the worst case scenario.

Figure 3 shows the dependence of the asymptotic rate on the dimension of the problem. As the number of dimensions increases, the asymptotic rate of convergence worsens but it is still much better than the worst rate.

Figure 4 shows the dependence of the asymptotic rate on the middle eigenvalues. Here we consider four-dimensional problems hence the second and third eigenvalues have been varied. When the middle eigenvalues are close to m or M the asymptotic rate is improved.

Figure 5 shows the attractors of the algorithm, plotted as a function of γ. It is clear from the graph where the algorithm exhibits chaotic behaviour. (b) depicts an example of the three-dimensional case with the middle eigenvalue being the midpoint between m and M, i.e. $\frac{m+M}{2}$. In any dimension, if the eigenvalues contain $\frac{m+M}{2}$, the bifurcation to chaos procedure starts at the smallest possible value of γ.

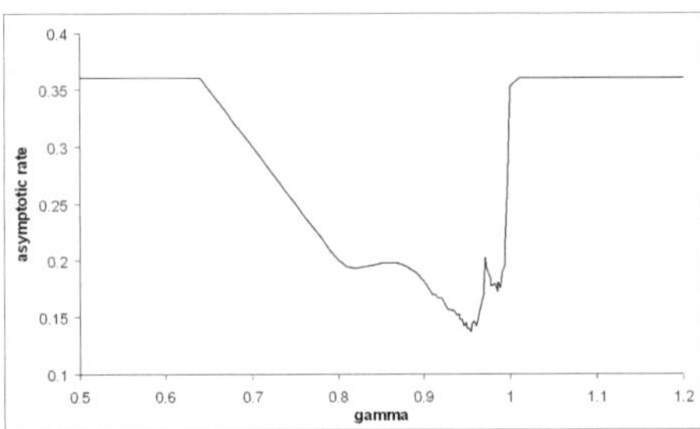

Figure 1. Asymptotic rate as a function of γ for $d = 4$ and eigenvalues $(1, 2, 3, 4)$.

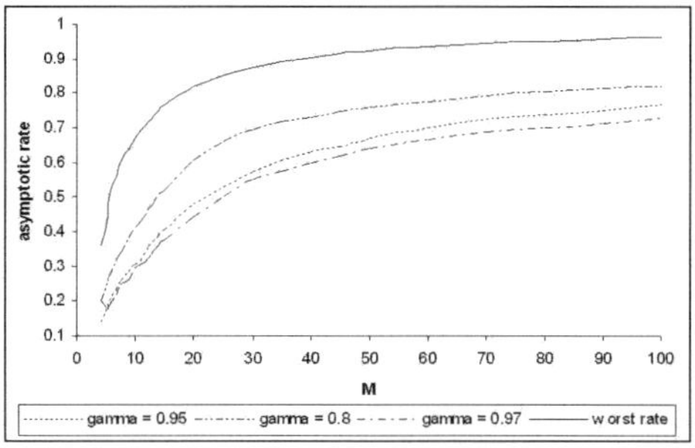

Figure 2. Asymptotic rate of convergence as a function of M for $d = 4$, eigenvalues are set to be $(m, m + \frac{M-m}{3}, m + \frac{2(M-m)}{3}, M)$.

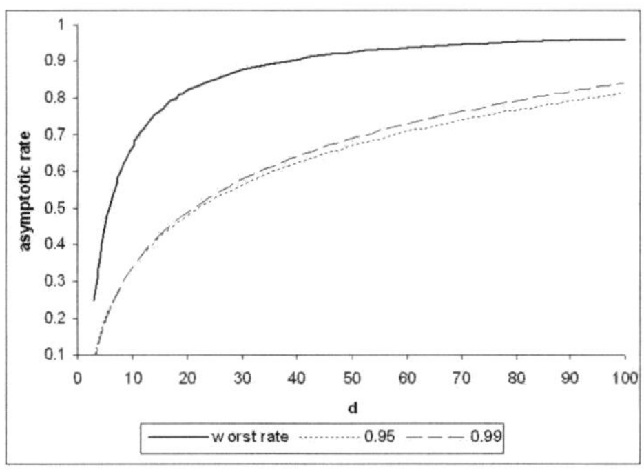

Figure 3. Asymptotic rate as a function of d for $\gamma = 0.95$ and $\gamma = 0.99$ with the eigenvalues evenly spaced.

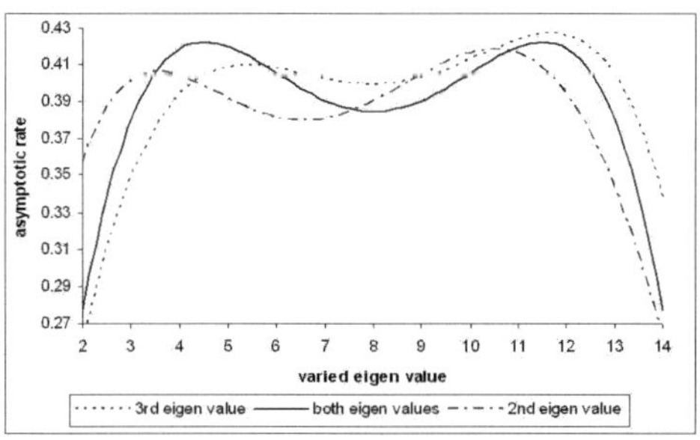

Figure 4. Asymptotic rate as a function of middle eigenvalues; $d = 4$, eigenvalues $(1, x, 14, 15)$, $(1, 2, x, 15)$ and $(1, x, 16 - x, 15)$.

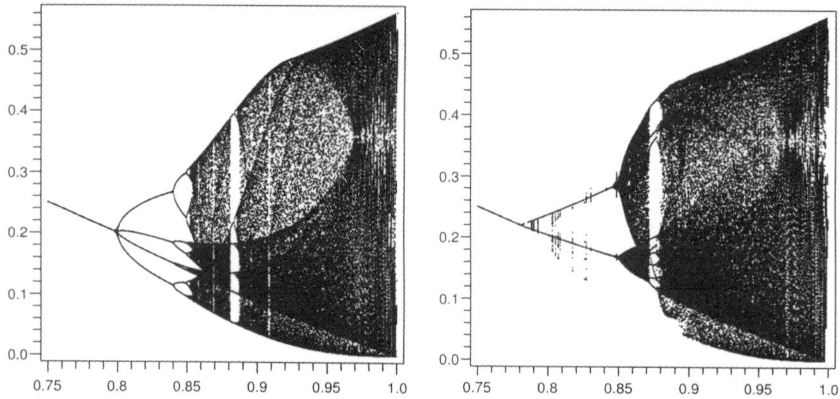

Figure 5. Attractors as a function of γ; (a) for $d = 2$ with eigenvalues $m = 1$ and $M = 4$, (b) for $d = 3$ with eigenvalues $(1, 2.5, 4)$.

5. Conclusion

In conclusion, the steepest descent algorithm can be greatly improved by the addition of γ, where γ is a constant. The optimal value of gamma appears to be slightly less than 1. This value of γ is outside the region where the algorithm converges to a single point and is contained within the section where chaotic behaviour is exhibited. The asymptotic rate of convergence of the algorithm also depends on the eigenvalues of the problem and is worst when the eigenvalues are large and evenly spaced.

References

1. D. G. Luenberger, *Linear and Nonlinear Programming*, 2nd edition, Addison-Wesley Publishing Company, Inc (1984).
2. L. V. Kantorovich and G. P. Akilov, *Functional Analysis*, 2nd edition, Pergamon Press, London (1982).
3. L. Pronzato, H. P. Wynn and A. Zhigljavsky, *Dynamical Search*, Chapman & Hall/CRC (2000).
4. L. Pronzato, H. P. Wynn and A. Zhigljavsky, Renormalised steepest descent in Hilbert space converges to a two-point attractor, *Acta Applicandae Mathematicae* **67**, 1–18 (2001).
5. L. Pronzato, H. P. Wynn and A. Zhigljavsky, Asymptotic behaviour of a family of gradient algorithms in $\mathbb{R}^d$ and Hilbert spaces, *Mathematical Programming*, to appear (2005).

A SYNERGY EXPLOITING EVOLUTIONARY APPROACH TO COMPLEX SCHEDULING PROBLEMS*

J. A. VÁZQUEZ RODRíGUEZ AND A. SALHI
Mathematical Sciences Department, The University of Essex, Wivenhoe Park Colchester, CO4 3SQ, U.K., E-mail: javazq@gmail.com, as@essex.ac.uk

We report on an innovative approach to solving Hybrid Flow Shop (HFS) scheduling problems through the combination of existing methods, most of which are simple heuristics. By judiciously combining these heuristics within an evolutionary framework, a higher level heuristic, a Hyper-Scheduler (HS), was devised. It was then, tested on a large array of HFS instances differing not only in input data, but crucially by the objective function used. The results suggest that HS success may well be due to it being successful at exploiting potential synergies between simple heuristics. These results are reported.

1. Introduction

A lot of research has been carried out on the design and implementation of algorithms for intractable scheduling problems with specific objectives. Although these efforts lead to relatively successful methods, the latter, due to their over-specialisation, are often ineffective when similar problems with different objectives were tackled. Moreover, often real world problems require that many objectives be considered at the same time, or that the same objective is allowed to change dynamically with time. In these cases, especially, existing methods leave a lot to be desired.[1,2] The present work is concerned with attempting to meet such demands, efficiently.

The term Hyper-Heuristic (HH) has been recently adopted[3,4] to refer to high level heuristics that coordinate the efforts of lower level ones. Instead of searching for a solution to the problem in hand, HH's search in the space of solution approaches (low level heuristics) for suitable ones for the problem in hand. These methods have been successfully applied to several practical problems.[5,6,7] In this paper, a Genetic Algorithm (GA) combined

*This work is supported by CONACYT grant 178473.

with a HH, into a Hyper-Scheduler (HS), is introduced and applied to Hybrid Flow Shop (HFS) scheduling problems. These problems are relatively unexplored, and even then most investigations consider a single objective function, namely minimising makespan.[8] Here, we consider HFS with other objective functions and combinations of these, giving problems with composite objective functions. The HS uses GA to solve part of the original problem, and, also, to find a combination of simple heuristics to finish off the solution. Note that HS is not a pure HH; it is more of a hybrid meta-heuristic (GA) and HH. The GA element schedules the first stage of the shop, but it is also used in the HH element to combine the simple heuristics in order to schedule the rest of the stages of the shop.

HS and several variants of the Single Stage Representation Genetic Algorithm[9] (SSRGA), were used to solve a large set of instances of the HFS problem. Note that SSRGA is a hybridisation of a GA with a low level heuristic (in this case a dispatching rule). The results show that, on the whole, HS performed better than its competitors, including the best SSRGA variant.

The rest of the paper is organised as follows. The next section presents a detailed description of HFS and the objective functions considered. Section 3 describes the proposed approach. Section 4 presents the details and results of the computational experiments. Section 5 is the conclusion.

2. Problem Definition

A HFS is a manufacturing environment in which a set of n jobs must pass through a series of m processing stages. At least one of the stages must have more than one identical machine in parallel.[10] HFS is a generalisation of the flow shop and the parallel machine environments, and is equally, NP-Hard.[11,12]

Let j represent a job, k a stage, and l a machine in a given stage. Let o_{jk} denote the operation of job j at stage k. The set of all operations to be processed at a given stage i.e. $\bigcup_{j=1}^{n} o_{jk}$, is O_k. The processing time required by o_{jk} is p_{jk}. Let r_{jk} be the release time of o_{jk}, i.e. the time when $o_{j,k-1}$ processing ends, or in the case of $k = 1$, the time when o_{j1} processing can start. The starting time of an operation is s_{jk} and its completion time c_{jk} $(c_{jk} = s_{jk} + p_{jk})$. The work remaining of an operation is denoted v_{jk}, i.e. $v_{jk} = \sum_{a=k}^{m} p_{ja}$.

Let A^{kl} be a set of operations $o_{jk} \in O_k$ assigned for processing to machine l in stage k. Let S^{kl} be a sequence of the elements in A^{kl} representing the order in which operations are to be processed. Let $S^k = \bigcup_{l=1}^{m_k} S^{kl}$ where m_k is the number of machines in stage k. S^k is a schedule for stage k because it represents the assignment and processing order of operations in it. The union of schedules of all stages is a full schedule, let us denote it S, i.e. $S = \bigcup_k S^k$. For S to be feasible the following must hold: $\bigcup_{l=1}^{m_k} A^{kl} = O_k$ $\forall k$ and $\bigcap_{l=1}^{m_k} A^{kl} = \emptyset$ $\forall k$. These constraints guarantee that all operations are assigned to strictly one processor. Let ψ be a HFS instance and Ω^ψ the set of all feasible schedules for ψ. The aim is to find a schedule $S \in \Omega^\psi$ such that its incurred cost $F_i(S)$ is minimum.

Let $F_i(S)$, $i = 1, 2, ..., 5$ be the set of objective functions of interest. These are: $F_1(S) = \sum w_j T_j + \sum w'_j E_j$, $F_2(S) = \{\max_j C_j, \sum w_j U_j\}$, $F_3(S) = \{\max_j C_j, \sum w_j W_j\}$, $F_4(S) = \{\sum C_j, \sum w_j E_j\}$, $F_5(S) = \{\sum w_j T_j, \sum w_j W_j\}$. C_j and w_j are the completion time and weight of job j, respectively. Let d_j be the due date of j, $T_j = \max(0, C_j - d_j)$ is the tardiness of j and $E_j = \max(0, d_j - C_j)$ its earliness. U_j is 1 if $C_j - d_j > 0$ and 0 otherwise, U_j is a penalty for late jobs. $W_j = C_j - s_{j1}$ is the waiting time of j in the shop. Note that W_j does not consider the waiting time in queue previous to the first stage of the shop.

Real world scenarios require to consider several criteria for decision making. For instance, the "Just in Time" and "lean" manufacturing philosophies require fast completion times, low inventory levels and to meet with the clients demands on time. F_1 to F_5 are inspired by these needs.

All the pairs of criteria involved in these functions are in conflict with each other. This justifies their inclusion in a single objective. However, there is the need for the Decision Maker (DM) to establish his/her preferences. The approach to handle this issue is described in Sec. 3.3.

3. Hyper-Scheduler

Exact methods, decomposition heuristics, methods exploiting bottleneck situations, adaptations of heuristics for the flow shop, and stochastic search methods have been suggested.[8,13,14] Four variants of the Single Stage Representation Genetic Algorithm (SSRGA) have been applied to HFS problems with different objective functions.[9] Each one of these variants combines GA, to schedule the first stage of the shop, and a simple dispatching rule (a different one in each SSRGA variant) to schedule the rest. It was observed that some of the SSRGA's were better at solving HFS with some objectives than HFS with others. There were, however, particular instances on which the best performing variant on the whole was not doing so well.

The interesting question we addressed here, is how to decide before-hand which SSRGA variant to use for a given instance of HFS. Furthermore, what are the benefits (if any) of combining several heuristics in a single SSRGA.

Several (13) simple heuristics were employed to generate a HH, to which we refer here as Hyper-Scheduler (HS). HS uses GA to search for a good permutation to schedule the first stage of the shop. Moreover, GA, is also used to search for a combination of the simple heuristics to schedule the rest of the stages of the shop. The same heuristics were also used to generate SSRGA variants. The rest of this section describes SSRGA and HS. Note that throughout this paper, low level heuristic and simple dispatching rule are interchangeable. A simple heuristic/dispatching rule consists of a selection criterion and an assignment procedure.

3.1. *Low Level Heuristics*

Each dispatching rule consists of three steps: (1) calculate the set of operations that are ready for processing at time t, (2) select one of them according to a selection criterion specific to the dispacthing rule, and (3) assign the operation to a given machine.

Let $O'_k \subseteq O_k$ be a set of operations that: (1) have not been assigned yet and (2) are ready to be processed at stage k (i.e. they have been released from the previous stage). Whenever a machine becomes idle, an operation $o_{jk} \in O'_k$ is selected according to one of the following simple heuristics criteria: the shortest r_{jk} (h_1), the shortest p_{jk} (h_2), the largest p_{jk} (h_3), the shortest $v_{jk} - d_j$ (h_4), the largest $v_{jk} - d_j$ (h_5), the shortest v_{jk} (h_6), the largest v_{jk} (h_7), the shortest $w_j p_{jk}$ (h_8), the largest $w_j p_{jk}$ (h_9), the shortest $w_j(v_{jk} - dj)$ (h_{10}), the largest $w_j(v_{jk} - d_j)$ (h_{11}), the shortest $w_j v_{jk}$ (h_{12}) or the largest $w_j v_{jk}$ (h_{13}). In the case that $O'_k = \emptyset$, o_{jk} will be the operation with the smallest release time. o_{jk} is assigned for processing after the last assigned operation to the first available machine in k. In all cases, ties are broken arbitrarily, and here, by preferring smallest job (j) or machine (l) indices.

3.2. *Solution Representation*

For the SSRGA, the adopted representation is a permutation $P = (p(1), p(2), \ldots, p(n))$ where every element $p(i)$ represents an operation to be scheduled at stage 1. Given a heuristic h_b to evaluate an individual P', operations are scheduled in the order $p'(1), p'(2), \ldots, p'(n)$ and assigned to the first idle machine at the first stage of the shop. The rest of the shop is scheduled according to h_b. A different SSRGA variant is obtained for each h_b. Call $h_b$ga the SSRGA variant which uses h_b to schedule stages 2 to m. The evaluation of an individual in SSRGA is as follows.

Algorithm EVALUATEINDIVIDUAL
input: P, F_i, h_b

1. $S = \emptyset$.
2. Generate S^1 according to P; set $S = S \bigcup S^1$.
3. For $k = 2, 3, ..., m$, generate S^k according to h_b; set $S = S \bigcup S^k$.
4. Return $F_i(S)$.

For HS, this representation was extended by adding to it an ordered set of heuristics HR containing $m - 1$ elements. Each element of HR is one of the heuristics already described, i.e. $HR_i \in HR \subseteq \{h_1, ..., h_{13}\}$. The ith heuristic in HR is the one to be used to prioritise the operations at stage $i + 1$. Example: $\{4, 2, 3, 1\}$, $\{h_4, h_9\}$ encodes a solution for a 4-job 3-stage shop. The operations in the first stage are considered for assignment in the order 4, 2, 3, 1. In stages 2 and 3, operations are scheduled in the order dictated by h_4 and h_9, respectively. In all stages, jobs are assigned to the first idle machine. The evaluation of individuals in the case of HS is as EVALUATEINDIVIDUAL with two modifications, (1) the algorithm takes as input a set of heuristics HR instead of a single heuristic h_b, and (2) at step 3, S^k is generated according to the $(k - 1)^{\text{th}}$ heuristic in HR.

3.3. *Fitness Evaluation*

Except for F_1, the criteria considered in the rest of the functions (F_2 to F_5) are measured in different and non-compatible units. For this reason, there is the need to manipulate their values so that they lie in the same range, in this case $[0, 1]$. To calculate fit_i, the fitness of an individual i, the following formula[15]

$$fit_i = \sum_{q=1}^{2} \lambda_q \frac{f_{qi} - min\_value_q}{max\_value_q - min\_value_q}, \tag{1}$$

is used, where λ_q such that $\sum_{q=1}^{2} \lambda_q = 1$, is the weight provided by the Decisions Maker (DM) as the priority for criterion f_q. f_{qi} is the objective value of q corresponding to individual i; min_value_q and max_value_q are the minimum and maximum values found so far for objective q. Note that Formula 1, is shown to be less sensitive than others, to the differences in ranges of the objective functions considered.[15]

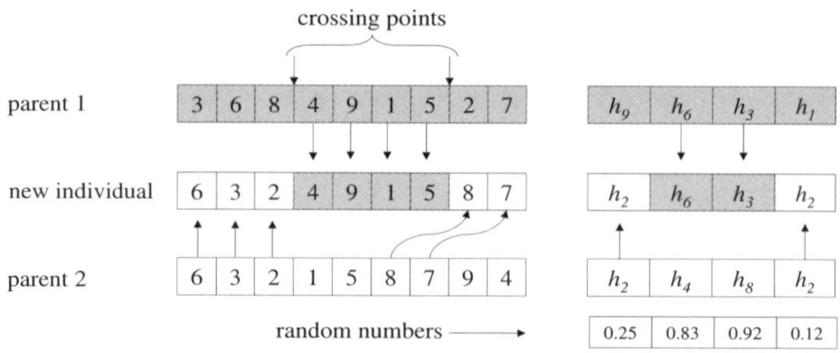

Figure 1. 2PX assuming that parent 1 is better fitted than parent 2.

3.4. *Genetic Operators and Parameter*

A *2-Point Crossover* (2PX), as shown in the left part of Fig. 1, was used to recombine P. In 2PX, two crossing points are randomly selected from parent 1 and the elements between them copied in the same position to the offspring. The missing jobs are copied from the second parent starting from the beginning of the permutation. In the case of HS the extra alleles representing the heuristics (HR) are selected randomly with a 0.7 probability from the fittest parent and 0.3 from the other (see right part of Fig. 1).

To mutate an individual, a random element is selected and moved into a randomly selected position in the sequence. The rest of the elements are moved to fill the released spaces as needed. In the case of HS, HR remains unchanged.

The following combination of operators and parameters was found to be appropriate when the search was limited to 10000 solution evaluations: a 90% crossover rate, a 9% mutation rate, 2 k tournament selection, 100 individuals in the population and 100 generations. 1% of the best individuals is retained in every generation.

4. Computational Experience

All SSRGA variants and HS were compared on a set of randomly generated instances minimising the objectives described in Sec. 2. All experiments were run on a 3.0 GHz processor with 1.0 GB of RAM running Windows XP. All implementations were in Java.

4.1. *Instance Generation*

We generated 128 shop configurations with the following parameter values: $n \in \{20, 40, 60, 80\}$; $m \in \{2, 4, 6, 8\}$; $m_k \in \{U\tilde{}(2,3), U\tilde{}(2,6)\}$; $p_{jk} \in \{U\tilde{}(50,70), U\tilde{}(10,100)\}$; $r_{jk} \in U\tilde{}(0, E(\sum_k p_{jk}))$; $d_{jk} \in \{U\tilde{}(0.9D, 1.1D), U\tilde{}(D, 1.5D)\}$, where $D = 1.5E(\sum_k p_{jk})$; $w_j \in \{U\tilde{}(2,8)\}$; $\lambda_1 \in U\tilde{}(0,1)$ and $\lambda_2 = 1 - \lambda_1$. These shop configurations were solved for the objective functions described in Sec. 2, for a total of 640 combinations each corresponding to an instance of HFS. Five runs were carried out with each algorithm on every problem instance. The best solution found for each instance by each algorithm was kept.

4.2. *Comparison Metrics*

Using the same arguments as in Sec. 3.3, the solutions generated by the heuristics for each instance were transformed into a single value (Q_H) which lies in the range $[0, 1]$. Formula 1 was adapted to this end as follows:

$$Q_H = \sum_{j=1}^{2} \lambda_q \frac{f_{qH} - min\_value_q}{max\_value_q - min\_value_q} \tag{2}$$

where f_{qH} is the value obtained by method H in f_q. In this case min_value_q and max_value_q are the minimum and maximum values, respectively, found by all methods. λ_q is as in Sec. 3.3. In the case of F_1 this formula is reduced to $(F_{1H} - \min\_value_{F_1})/(max\_value_{F_1} - min\_value_{F_1})$.

Q_H is an appropriate metric because it measures the performance of every heuristic in relation to the rest and the DM's preferences provided in terms of priorities λ_1 and λ_2, see Sec. 3.3. Next, the mean Q_H value of each of the heuristics on the 5 sets of instances is presented. The standard deviation on this value is presented as a measure of the stability of the methods. Finally, the number of times that every algorithm achieved a $Q_H \leq 0.1$, which we consider as a success, is also presented.

4.3. *Results*

By looking at Table 1, it can be appreciated that the best performing heuristic is HS. Except for F_3 (on which it was the second best) it obtained the best Q_H mean value on all instances. Furthermore, HS outperformed the best among the SSRGA's on each particular objective function. Moreover, as shown in Table 2, HS has, also, on the whole, the lowest standard deviation on the Q_H value, which is a good indicator of robustness. Finally,

Table 1. Q value of each heuristic on the 128 instances for each of 5 different composite objective functions.

	F_1	F_2	F_3	F_4	F_5	Total
h_1ga	0.230	0.334	0.296	0.277	0.285	0.284
h_2ga	0.343	0.427	0.358	0.253	0.298	0.336
h_3ga	0.579	0.430	0.455	0.480	0.479	0.485
h_4ga	0.487	0.468	0.381	0.332	0.424	0.418
h_5ga	0.415	0.379	0.324	0.377	0.387	0.376
h_6ga	0.359	0.527	0.470	0.270	0.338	0.393
h_7ga	0.547	0.355	0.329	0.427	0.428	0.417
h_8ga	0.634	0.423	0.465	0.296	0.574	0.478
h_9ga	0.388	0.424	0.287	0.372	0.287	0.352
$h_{10}ga$	0.235	0.395	0.315	0.318	0.234	0.299
$h_{11}ga$	0.730	0.416	0.486	0.358	0.637	0.525
$h_{12}ga$	0.679	0.472	0.532	0.318	0.606	0.521
$h_{13}ga$	0.318	0.377	0.222	0.354	0.273	0.309
HS	0.175	0.275	0.238	0.244	0.228	0.232

Table 2. Standard deviation from the Q value of each heuristic on the 128 instances for each of 5 different composite objective functions.

	F_1	F_2	F_3	F_4	F_5	Total
h_1ga	0.286	0.265	0.217	0.241	0.237	0.252
h_2ga	0.269	0.266	0.208	0.232	0.179	0.240
h_3ga	0.316	0.284	0.294	0.310	0.250	0.295
h_4ga	0.265	0.290	0.237	0.236	0.229	0.258
h_5ga	0.228	0.281	0.213	0.261	0.232	0.245
h_6ga	0.270	0.288	0.271	0.235	0.210	0.272
h_7ga	0.305	0.269	0.250	0.283	0.231	0.278
h_8ga	0.305	0.252	0.259	0.218	0.307	0.294
h_9ga	0.330	0.257	0.209	0.258	0.234	0.266
$h_{10}ga$	0.291	0.269	0.259	0.237	0.246	0.267
$h_{11}ga$	0.319	0.255	0.278	0.244	0.332	0.319
$h_{12}ga$	0.285	0.271	0.275	0.235	0.307	0.301
$h_{13}ga$	0.316	0.274	0.206	0.248	0.242	0.265
HS	0.246	0.229	0.231	0.243	0.236	0.239

Table 3 presents the number of times that the Q_H value achieved by the heuristics was ≥ 0.1 In this case, even on F_3, HS is the superior solver.

One interesting interpretation of these results is that they provide evidence for the positive effect of the synergy of simple heuristics, brought up by HS. This synergy may well be calculated explicitly for different combinations of simple heuristics to allow for a more effective and efficient choice of solution approach, given a HFS specification, objective function, dimensions and other characteristics.

Table 3. Number of times that each heuristic obtained a Q value ≤ 0.1 on the 128 instances for each of 5 different composite objective functions.

	F_1	F_2	F_3	F_4	F_5	Total
$h_1$ga	63	29	27	43	37	199
$h_2$ga	26	15	14	51	15	121
$h_3$ga	11	24	22	19	10	86
$h_4$ga	5	17	21	33	9	85
$h_5$ga	10	27	21	31	18	107
$h_6$ga	19	10	11	45	18	103
$h_7$ga	12	32	33	24	11	112
$h_8$ga	5	20	12	36	12	85
$h_9$ga	42	14	31	30	36	153
h_{10}ga	63	20	35	37	50	205
h_{11}ga	10	17	11	30	11	79
h_{12}ga	5	18	10	34	10	77
h_{13}ga	48	23	47	31	43	192
HS	75	38	52	49	55	269

5. Conclusion

A solution approach, the Hyper Scheduler (HS), to HFS with composite objective functions was presented and compared with 13 variants of SS-RGA. HS showed a better average performance, and a higher success rate (as defined in Sec. 4.2) than its competitors.

The results suggest that HS exploits the synergy of simple heuristics when combined. It is, perhaps, this synergy which explains the success rate of HS, and helps the search to find high quality solutions. Better performing algorithms may be obtained by exploring alternative ways of exploiting the potential synergy between individual heuristics.

HS seems to be a suitable base for an algorithm to approach multi-objective scheduling other than HFS. Further work is being carried out to improve HS, and to adapt it to generate the set of efficient solutions on the pareto frontier in a multi-objective scheduling context.

References

1. K. McKay, M. Pinedo and S. Webster, Practice-focused research issues for scheduling systems, *Production and Operations Management* **11**, 249–258 (2002).
2. S. F. Smith, Is scheduling a solved problem? In: G. Kendall, E. K. Burke, S. Petrovic and M. Gendreau (eds.), *Multidisciplinary Scheduling: Theory and Applications*, Springer, pp. 3–17 (2003).
3. P. Cowling, G. Kendall and E. Soubeiga, A hyperheuristic approach to

scheduling a sales summit, In: E. K. Burke and W. Erben (eds.), *LNCS 2079, Practice and Theory of Automated Timetabling III : Third International Conference, PATAT 2000*, Springer-Verlag, pp. 176–190 (2000).
4. E. Burke, E. Hart, G. Kendall, J. Newall, P. Ross and S. Schulenburg, Hyper-heuristics: An emerging direction in modern search technology, In: F. Glover and G. Kochenberger (eds.), *Handbook of Metaheuristics*, Springer, pp. 457–474 (2003).
5. H. Terashima-Marín, A. Morán-Saavedra and P. Ross, Forming hyper-heuristics with GAs when solving 2d-regular cutting stock problems, In: *Proceedings of Congress on Evolutionary Computation CEC(2005)*, IEEE Press, pp. 1104–1110 (2005).
6. G. Kendall and M. Mohamad, Channel assignment optimisation using a hyper-heuristic, In: *Proceedings of the 2004 IEEE Conference on Cybernetic and Intelligent Systems (CIS2004)*, IEEE, pp. 790–795 (2004).
7. K. Chakhlevitch and P. Cowling, Choosing the fittest subset of low level heuristics in a hyper-heuristic framework, In: G. Raidl and J. Gottlieb (eds.), *EvoCOP 2005, LNCS 3448*, Springer-Verlag, Berlin Heidelbergh, pp. 23–33 (2005).
8. H. Wang, Flexible flow shop scheduling: Optimum, heuristics and artifical intelligence solutions, *Expert Systems* **22**, 78–85 (2005).
9. J. A. V. Rodríguez and A. Salhi, Performance of single stage representation genetic algorithms in scheduling flexible flow shops, In: *Congress on Evolutionary Computation (CEC2005)*, IEEE Press, pp. 1364–1371 (2005).
10. S. A. Brah, *Scheduling in a Flow Shop with Multiple Processors*, PhD thesis, University of Houston (1988).
11. J. N. D. Gupta, Two-stage hybrid flow shop scheduling problem, *Operational Research Society* **39**, 359–364 (1988).
12. J. A. Hoogeveen, J. K. Lenstra and B. Veltman, Preemptive scheduling in a two-stage multiprocessor flow shop is NP-hard, *European Journal of Operational Research* **89**, 172–175 (1996).
13. R. Linn and W. Zhang, Hybrid flow shop scheduling: A survey, *Computers & Industrial Engineering* **37**, 57–61 (1999).
14. A. Vignier, J. C. Billaut and C. Proust, Les problèmes d'ordonnancement de type flow-shop hybride: état de l'art. *Operations Research* **33** 117–183 (1999).
15. P. J. Bentley and J. P. Wakefield, Finding acceptable solutions in the pareto-optimal range using multiobjective genetic algorithms, In: P. K. Chawdhry, R. Roy and R. K. Pant (eds.), *Soft Computing in Engineering Design and Manufacturing*, Part 5, Springer Verlag London Limited, pp. 231–240 (1997).

OPTIMAL CONFIGURATION, DESIGN AND OPERATION OF HYBRID BATCH DISTILLATION/PERVAPORATION PROCESSES

T. M. BARAKAT AND E. SØRENSEN*

Centre for Process Systems Engineering, Department of Chemical Engineering University College London, Torrington Place, London WC1E 7JE, U. K.

This paper considers for the first time, the simultaneous optimisation of configuration, design and operation of hybrid batch distillation/pervaporation processes by considering all possible process structures. The overall problem is formulated as a mixed integer dynamic optimisation (MIDO) problem. The optimisation strategy comprises of an overall economics index that encompasses capital investment, operating costs and production revenues. Furthermore, rigorous dynamic models developed from first principles for distillation and pervaporation are used. A case study for the separation of homogeneous tangent-pinch (acetone-water) mixtures is presented. It is found that fully integrated hybrid configuration is economically favourable when compared to a conventional distillation process, however, this configuration may be more complex to operate and control.

1. Introduction

Distillation is the most commonly used technique for separating liquid mixtures within the chemical industries despite being an energy and capital intensive process. Many mixtures commonly encountered in the fine chemical and pharmaceutical industries are, however, difficult or impossible to separate by normal distillation due to azeotropic behaviour, tangent pinch or low relative volatilities. Pervaporation has been hailed as an alternative to distillation for such mixtures as the separation mechanism is different, relying on differences in solubility and diffusivity between the components in the mixture and not vapour-liquid equilibrium as in distillation. Recently, hybrid processes have been proposed where a distillation column unit and a pervaporation unit are integrated into one process. In such a process, the shortcomings of one method are outweighed by the benefits of the other, allowing for significant savings in terms of energy consumption and cost. Although the optimisation of design and operation of continuous hybrid distillation/pervaporation processes has been

* Author to whom correspondence should be addressed. E-mail: e.sorensen@ucl.ac.uk

attempted before, this is the first time the simultaneous optimisation of *configuration*, design, and operation is considered and it is in this work applied to *batch* hybrid distillation/pervaporation processes.

The two units can be integrated in different ways; the pervaporation unit can be positioned before the distillation column, after the column, or fully integrated. Depending on the particular separation task, the configuration, design and operation of a hybrid should be optimised to achieve the most suitable performance.

Eliceche *et al.*[1] carried out optimisation studies of operating conditions for a continuous hybrid distillation/pervaporation system consisting of an azetropic distillation column connected via a side stream to a pervaporation unit. They solved the optimisation problem by minimising the operating costs, however, they did not consider the design or configuration of the hybrid system. Szitkai *et al.*[2] optimised the design and operation of a continuous hybrid dehydration system using an MINLP model to minimise the annual operating costs of a single, post-distillation, hybrid configuration. Recently, Kookos[3] proposed a methodology for the structural and parametric optimisation of continuous hybrid separation systems. He described the superstructure of the hybrid process using a simplified steady-state mathematical model where it was assumed that all streams taken from, or returned to, the distillation column were vapour streams. The methodology is therefore not suitable for other membrane processes, such as pervaporation, or for dynamic systems, such as batch processes.

Most separations within the fine chemical or pharmaceutical industries are run batch-wise. The optimal design and operation of batch distillation columns has received considerable interest in recent years, particularly, in terms of novel column configurations such as inverted, middle vessel and multivessel column configurations Low and Sørensen.[4] Adding a pervaporation unit to the system, either before, after or fully integrated, adds complexity to the system but also more degrees of freedom which, if properly chosen, can further increase the profitability of the system, particularly for difficult separations such as that of azeotropic mixtures.[5]

The design engineer is thus faced with a difficult task: to determine not only the best design and operation of the separation process, but also which separation technique to use and, if considering a hybrid system, how the two units should be combined. The objective of this work is to propose an optimal process synthesis procedure that allows the determination of the optimal process type, its configuration and design for a given separation duty. This procedure can be extended for any number of separation process alternatives, but the discussion in this work will be limited to batch distillation, batch pervaporation and batch hybrid distillation processes. In the next section, the batch separation synthesis problem by utilising a process superstructure is presented, followed by the objective function formulation and optimisation problem definition. The mathematical models used in this study are then presented as well as an overview of the optimisation strategy. Finally, the batch separation process problem optimisation and its solution strategy are applied to a case study for the separation of a tangent-pinch mixture (acetone-water).

2. The Batch Separation Synthesis Problem

2.1. *Problem Definition*

The objective of batch separation process synthesis is to determine the optimal separation process which results in the most economical benefit when processing a given separation task. To achieve this objective, optimal configuration, design and operation must be considered *simultaneously* based on an objective function that encapsulates capital investment, operating costs and production revenues.

There is a trade-off between capital investment in terms of equipment and performance and also between operational decisions and performance. When considering a batch distillation column for instance, it is possible to design the column with a low number of trays operating at high reflux ratio, or alternatively, design the column with more trays and operating at lower reflux ratio and still achieve the same separation requirements. The decision will, however, clearly have an impact on the profitability of the process.

2.2. *Superstructure*

The optimal synthesis of batch separation processes superstructure is presented next. The superstructure (Fig. 1) incorporates three separation processes: batch distillation, batch pervaporation and hybrid batch distillation/pervaporation processes. The superstructure proposed here allows (for a given set of binary and integer variables) not only the most economical process to be selected but also its optimal operation and design to be determined in order to carry the required separation duty optimally. Similar work on hybrid processes superstructure has been proposed by Kookos[3] but it only allows for the optimisation of the hybrid process and therefore exploring either distillation or pervaporation as a potential separation process is not possible. The pervaporation membrane separation stage used in this superstructure, as shown in Fig. 1, consists of a number of identical membrane modules (N_m) connected in parallel.[6] The membrane stage feed stream is assumed to be distributed evenly between the membrane modules and

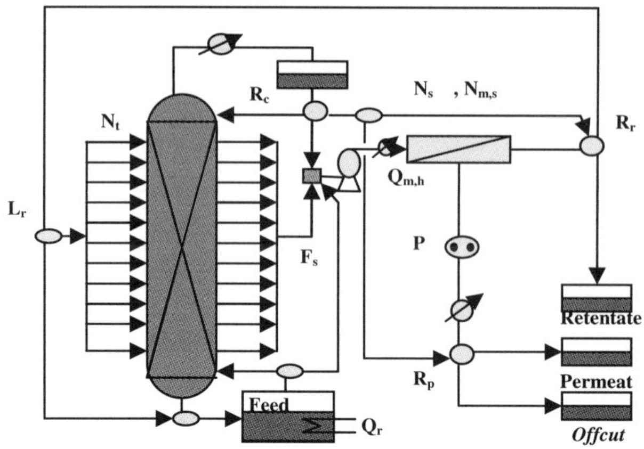

Figure 1. Batch separation superstructure.

therefore a single mathematical model can be used to describe the modules. This method proposed by Marriott and Sørensen[6] was found to reduce the computational time significantly.

A rigorous distillation column tray model is employed. Each tray is modelled to accommodate for two extra potential streams in addition to the regular vapour and liquid inlet/outlet to the neighbouring trays. The first stream is a side draw stream *to* the pervaporation unit in a hybrid configuration if the tray is selected as a membrane feed tray. The second stream is an inlet stream *from* the pervaporation unit in a hybrid configuration if the tray is selected as a retentate recycle tray.

2.3. *Objective Function*

The optimal design and operation of batch separation processes, as it is considered in this paper, is determined as the most economical process design and corresponding operating policy that will satisfy all specified separation requirements and constraints. The optimal solution is a trade-off between capital and operating costs versus production revenue, and is reflected in the formulation of the objective function as shown below:

$$P_A = \left(\frac{\sum_{i=1}^{Nc} C_i M_{i,f} - C_{feed} M_{feed}}{t_f + t_s} \right) \times T_A - ACC - AOC \tag{1}$$

The annualised capital costs and operating costs for the batch distillation column[4] is given by:

$$ACC_c = K_1 N_t^{\,0.802} V^{0.533} + K_2 V^{0.65} \tag{2}$$

$$AOC_c = C_{ut} \times \left(Q_{reb} + Q_{c,cond}\right) \tag{3}$$

The annualised capital costs and operating costs for the batch membrane process is given by:

$$ACC_m = ACC_m + ACC_{m,anc} \tag{4}$$

$$AOC_m = C_{ut} \times \left(Q_{m,h} + Q_{m,cond}\right) + AOC_{m,p} + AOC_{m,t} \tag{5}$$

The annualised capital costs and operating costs for the hybrid batch distillation column is given by:

$$ACC_{hyb} = ACC_c + ACC_m \tag{6}$$

$$AOC_{hyb} = AOC_c + AOC_m \tag{7}$$

2.4. *Optimisation Problem Formulation*

The objective of the synthesis procedure is to maximise the profitability defined by the objective function above, subject to process type, process model equations and all separation duty constraints. The optimisation problem is therefore:

Given a mixture M_{feed} with number of components N_C to be separated, minimum product purities $x_i^{\min}$, minimum product recoveries $M_{i,f}$, price structure of feed and products, C_{feed} & C_i, total production time available per annum T_A; determine the optimum set of design variables u_d, and the optimum set of operation variables u_o, to achieve the maximum objective function value P_A (1):

$$Max_{u_d,u_o} P_A$$

subject to:

$$f(\dot{x}, x, t, u_d, u_o) = 0 \tag{8}$$

$$x_i(t_f) \ge x_i^{\min},\ \forall i = 1,\ldots,N_C \tag{9}$$

$$u_d^{\min} \le u_d \le u_d^{\max} \tag{10}$$

$$u_o^{\min} \leq u_o \leq u_o^{\max} \tag{11}$$

Equation (8) represents the mathematical process model of the batch separation process; x is a vector of process state variables, u_d and u_o denote the vectors of design and operating control variables, respectively. Equation (9) represents the product purity constraints imposed which must be satisfied at the end of the batch. Equations (10) and (11) represent the physical and optimisation bounds of the design and operating control variables, respectively.

For the batch distillation process, the set of operating variables include vapour boilup rate and column reflux ratio profile, *i.e.* $u_o^c = \{V, R_C\}$. The vapour boilup rate can subsequently be used to determine the diameter of the column (*e.g.* using Guthrie's correlation, $D \propto \sqrt{V}$) as well as the reboiler and condenser heat loads. Design variables include the optimal number of trays N_t, *i.e.* $u_d^c = \{N_t\}$.

For the batch pervaporation process, the set of operating variables include retentate recycle ratio R_r, permeate pressure P and feed tank heat load $Q_{m,h}$, *i.e.* $u_o^m = \{R_r, R_p, Q_{m,h}\}$. The set of design variables include number of membrane modules N_m, *i.e.* $u_d^m = \{N_m\}$. For the hybrid batch distillation process, the set of operating variables and design variables are a combination of the previous two processes with an additional design variable for the retentate recycle location L_r.

2.5. *Process Models*

The distillation model is based on the approach of Low and Sørensen[4] which disposes of some of the common modelling assumptions, such as negligible tray holdup and constant molal overflow that may otherwise have a significant impact on the optimal solution. The main features of the model are: dynamic mass and energy balances and rigorous thermodynamics through the use of liquid and vapour fugacities. The assumptions retained in this work include no entrainment effects, no downcomer dynamics, adiabatic column operation, phase equilibrium and perfect mixing.

The mathematical model used in this study to describe the performance of hollow fibre pervaporation membrane modules is similar to that of Marriott and Sørensen.[6] The model features a 1-D plug flow pattern through the membrane fibres and module shell. Furthermore, dynamic mass and energy balances, as well as rigorous thermodynamics have been included. The membrane characterisation equations are from Tsuyumoto *et al.*[7] Concentration variation perpendicular to the bulk flow direction is neglected and prefect mixing throughout is also assumed.

The mathematical model of the hybrid distillation/pervaporation is a combination of the distillation and pervaporation models outlined above. It should be noted that the approach outlined in the following can be used with models of any modelling complexity although the confidence in the results will depend on the accuracy of the models.

2.6. *Solution Methodology*

Simultaneous consideration of optimal design and operation of batch separation processes translates into an optimisation problem with both discrete (*e.g.* number of trays and number of membrane modules) and continuous variables (*e.g.* reflux and recycle ratios). Furthermore, the nonlinear dynamic models used here as well as the nonlinear objective function defined, transforms the problem into a complex mixed integer dynamic optimisation (MIDO) problem. MIDO problems are difficult to solve using conventional optimisation techniques due to the high nonconvexity and complex search space topography nature of these problems and the combination of integer and continuous variables, and there is much ongoing research on developing robust and practical solution algorithms.[4] The proposed batch distillation superstructure is solved using a genetic algorithm (GA) optimisation framework that works through the conventional genetic algorithm operators of which further details can be found in Goldberg.[8]

In this work, a given solution set consists of all decision variables are represented in the genome as direct real values instead of converted binary bits and mapping which has been found to be less efficient.[9] The initial population is created randomly. The objectives and constraints of each individual in this population are evaluated using the *gPROMS* simulation software.[10] A penalty function procedure is applied as described by Low & Sorensen[4] when necessary to encourage the GA to drive the population towards feasibility. Solutions are assigned a fitness score based on annual profitability of each genome.

The GA procedure uses a roulette selection scheme, 75% replacement rate, 75% crossover rate, 10% mutation rate and a stopping criterion based on the number of generations. A steady-state population strategy is employed as described in Low & Sorensen.[4] The procedure has been implemented using the GALib genetic algorithm library.[11]

2.7. *Results and Discussion*

The optimal process synthesis procedure developed in this work is demonstrated by considering the separation of a tangent-pinch mixture of acetone and water. The separation process specifications are shown in Table 1.

The batch process operation is separated into three task intervals, the first period is for total reflux/recycle followed by a water product collection period, while the acetone product is purified to the required purity, and finally an acetone product collection period.

2.8. *Optimal Solution*

The optimum solution sets of the superstructure and that of a comparative batch distillation case are shown in Table 2. A fully integrated hybrid distillation/Pervaporation process was the optimal process solution. The optimal number of trays and membrane modules are found to be *30 & 2*, respectively. Optimal batch time is found to be *104.53 minutes* with interval split *81.35, 15.0 & 8.18 minutes*. Corresponding reflux and recycle ratios are found in Table 2. Optimal reboiler vapour load is found to be 4.94 *mol/s* with optimal sidedraw flowrate of *2.72 mol/s* . The optimal membrane inlet heater temperature is found to be *330 K* and permeate side pressure is *300 Pa*. Optimal sidedraw location is found to be at tray 2, with optimal retentate return location at tray *5* of the hybrid column, where tray one is at the top of the column.

The optimal design and operation of batch hybrid distillation/pervaporation is found to be the most profitable process alternative that meets all separation requirements for this case study (see Table 2), with an estimated profit of 18.07 M£ per annum. The optimum process is found to be 26% more profitable than an optimised batch distillation process as shown in Table 2. A more comprehensive

Table 1. Unit specifications and operating conditions.

Property	Value	Property	Value
Feed composition, $x_{i,feed}$ (*mol fraction*)		Available production time T_A (*hr*)	7920
Acetone	0.70	Batch setup time, t_s (*min*)	30
Water	0.30	Cost, C_i (*£/mol*)	
Batch size, M_{feed} (*mol*)	20,000	Acetone	0.606
Tray/Condenser holdup (*mol*)	0.1	Water	0.040
Product purities, $x_{i,f}$ (*mol frac.*)	≥ 0.97	Feed	0.150
Product recoveries, $M_{i,f}$	≥ 0.70	Offcut	0.010
		Utility (£/MJ)	0.019

study of how the other processes, batch distillation and batch Pervaporation, compare with hybrids can be found in Barakat & Sorensen.[12]

3. Conclusions

In this work, the optimal synthesis of batch separation processes has been considered. The synthesis problem is solved through simultaneous consideration of optimal design and corresponding operating policy of all process alternatives by utilising a process superstructure. The optimal solution is defined as the most economical process configuration, design and operation that achieves all separation requirements. The problem objective function reflects the various trade-offs between design and operation decision variables versus production revenue as well as that of capital investments versus operating costs.

Hybrid batch distillation configuration was found to be the optimal synthesis solution for the separation of tangent-pinch acetone-water case considered, this was further verified by comparison to an optimised batch distillation process. The proposed methodology can be extended to allow for the synthesis of any number of separation alternatives by incorporating them into a single process superstructure. However, as alternatives increase, the required computational time to solve such a superstructure can also increase significantly.

Table 2. Optimal solutions sets.

Optimal Variables Set

$$u_{d,c} = \{N_t\} = \{30\}$$

$$u_{o,c} = \{t_f, t_i, R_c, V\} = \left\{117.8, \begin{bmatrix}0.02\\117.0\\0.06\end{bmatrix}, \begin{bmatrix}0.75\\0.57\\0.71\end{bmatrix}, 4.70\right\}$$

$$u_{d.hyb} = \{N_t, N_m, F_s, L_r\} = \{30,2,2,5\}$$

$$u_{hyb} = \{t_f, t_i, R_c, R_r, R_p, P, T_o, V, F_{side}\}$$

$$= \left\{104.5, \begin{bmatrix}81.3\\15.0\\8.1\end{bmatrix}, \begin{bmatrix}1.0\\1.0\\1.0\end{bmatrix}, \begin{bmatrix}0.16\\0.16\\0.15\end{bmatrix}, \begin{bmatrix}0.0\\0.0\\0.0\end{bmatrix}, 300, 330, 4.94, 2.72\right\}$$

Annual Profit (/year),	***Column***	***Hybrid***
	£17,770,000	***£ 19,030,000***

Nomenclature

ACC	Annualised equip. capital cost (£/yr)	t_f	Total processing time (min)
AOC	Annualised equip. operating costs (£/yr)	t_s	Setup time (min)
C_i	Selling price of product i (£/mol)	T_A	Production time available per annum
C_{feed}	Cost price of feed (£/mol)	u_d	Vector of design variables
C_{ut}	Utilities cost (£/MJ)	u_o	Vector of operation variables
F_{feed}	Membrane feed flowrate (mol/s)	V	Column boilup rate (mol/sec)
F_s	Location of the column sidedraw	x	Vector of state variables
F_{side}	Flowrate of the sidedraw stream	x_i	Conc. of i in mixture
K_1	Guthrie's coeff. for column shell cost	x_i^{min}	Minimum conc. of i in mixture
K_2	Guthrie's coeff. for exchangers cost		
L_r	Retentate recycle location	**Super**	
M_{feed}	Batch size (mol)	c	Column
$M_{i,f}$	Final product i recovery	m	Membrane
N_c	Number of components	**Sub**	
N_m	Number of membrane modules	anc	Ancillary
N_t	Number of column trays	c	Column
P_A	Annual profit (£/yr)	cond	Condenser
P	Permeate pressure (Pa)	m	Membrane
Q	Heat load (kW)	reb	Reboiler
$Q_{m,h}$	Pervaporation heat load (kW)	m,h	Membrane system feed heater
R_c	Column internal reflux ratio	m,t	Membrane system turbine
R_p	Permeate offcut ratio	m,p	Membrane system feed pump
R_r	Retentate recycle ratio	hyb	Hybrid system

References

1. A. Eliceche, M. C. Daviou, P. M. Hoch and I. Ortiz Uribe, *Comp. & Chem. Eng.* **26**(4), 563-573 (2002).
2. Z. Szitkai, Z. Lelkes, E. Rev and Z. Fonyo, *Chem. Eng. and Proc.* **41**(7), 631-646 (2002).
3. I. K. Kookos, *Ind. Eng. Chem. Res.* **42**(8), 1731-1738 (2003).
4. K. H. Low and E. Sørensen, *AIChE J.* **49**(10), 2564-2576 (2003).
5. V. Van Hoof, L. Abeele, A. Buekenhoudt, C. Dotremont and R. Leysen, *Sep. and Pur. Tech.* **37** (1), 33-49 (2004).
6. J. I. Marriott and E. Sørensen, *Chem. Eng. Sci.* **58**(22), 4975-4990 (2003).
7. M. Tsuyumoto, A. Teramoto and P. Meares, *Journal of Membrane Science* **13** (1)**,** 83-94 (1997).
8. D. E. Goldberg, Addison-Wesley, Boston, London (1989).
9. D. Coley, World Scientific Publishing, Singapore, 1st ed. (1999).
10. Process Systems Enterprise Ltd., User's Manual, UK (2005).
11. M. Wall, *GAlib: C++ Library of Genetic Algorithm Components, version 2.4.5*, (1999) http://lancet.mit.edu./ga.
12. T. Barakat and E. Sorensen, In: *proceedings of the 7th World Congress of Chemical Engineering*, Glasgow (2005).

OPTIMAL ESTIMATION OF PARAMETERS IN MARKET RESEARCH MODELS

V. SAVANI

Department of Mathematics, Cardiff University
Cardiff, CF24 4AG, U.K., E-mail: SavaniV@cardiff.ac.uk

In the modeling of market research data the so-called Gamma-Poisson model is very popular. The model fits the number of purchases of an individual product made by a random consumer. The model presumes that the number of purchases made by random households, in any time interval, follows the negative binomial distribution. The fitting of the Gamma-Poisson model requires the estimation of the mean m and shape parameter k of the negative binomial distribution. Little is known about the optimal estimation of parameters of the Gamma-Poisson model. The primary aim of this paper is to investigate the efficient estimation of these parameters.
Keywords: Gamma-Poisson model, market research, maximum likelihood, moment estimators, negative binomial distribution.

1. Introduction

The Gamma-Poisson process has been successfully applied in the modeling of, for example, accidents and sickness, [1] market research,[2] risk theory[3] and clinical trials.[4] The Gamma-Poisson process implies that data observed over any time interval follows the negative binomial distribution (NBD).

The fitting of mixed Poisson processes to observed data in literature[2] has mainly focussed on the fitting of the NBD when considering data observed over fixed time intervals. Fisher[5] and Haldane[6] independently considered estimating the NBD parameters using the maximum likelihood (ML) approach. As an alternative, simple moment based estimation methods have been considered.[7,8,9] Moment based estimators have been developed since maximum likelihood estimators are sometimes impractical.

In this paper it will be shown that the efficiency of the moment based estimation methods depend on the time interval over which data is observed. Additionally, depending on the moment based method used, it is not necessarily the case that the largest time interval should be taken to obtain the most efficient estimator. This is practically important. For example, in the case of market research, consumers buying behavior may be observed for any arbitrary length of time, and the NBD fitted to the observed data. However, there is no indication as to how long data should be observed for, in order to obtain efficient parameter estimates.

2. Background

The Gamma-Poisson Process

The most general form of the Gamma-Poisson process was noted by Grandell[3] who considered the Gamma-Poisson process as a mixed Poisson process. Let $\mathbf{X} = \{X(t_1), X(t_2), \ldots, X(t_n)\}$ be a random vector, $\mathbf{x} = \{x_1, x_2, \ldots, x_n\}$ with $0 = x_0 \leqslant x_1 \leqslant x_2 \leqslant \ldots \leqslant x_n$ and let $0 = t_0 \leqslant t_1 \leqslant \ldots \leqslant t_n$ represent an increasing sequence of time points where n is a positive integer, then given parameters $a > 0$ and $k > 0$, the finite dimensional distribution of the Gamma-Poisson process is

$$\mathbb{P}(\mathbf{X} = \mathbf{x}) = \frac{\Gamma(k + x_n)}{\Gamma(k)} \left(\prod_{i=0}^{n-1} \frac{(t_{i+1} - t_i)^{x_{i+1} - x_i}}{(x_{i+1} - x_i)!} \right) \frac{a^{x_n}}{(1 + at_n)^{x_n + k}}.$$

The Negative Binomial Distribution

Consider the finite dimensional distribution of the multivariate Gamma-Poisson process in the case where $n = 1$ and $t_0 = 0$ then

$$p_{x_1} = \mathbb{P}(X(1) = x_1) = \frac{\Gamma(k + x_1)}{\Gamma(k) x_1!} \left(\frac{1}{1 + at_1} \right)^k \left(\frac{at_1}{1 + at_1} \right)^{x_1}. \tag{1}$$

The one dimensional distribution of the Gamma-Poisson process is the NBD with parameters (at_1, k). The parameter a is a scale parameter of the distribution, so without loss of generality we may consider the parameterization (a, k) instead of (at_1, k). The NBD can be re-parameterized by (m, k), where $m = ak$ denotes the mean of the distribution. Anscombe[7] noted that the maximum likelihood and all natural moment based estimators for (m, k) are asymptotically uncorrelated for an i.i.d. NBD sample. The estimation of NBD parameters in literature has therefore only focussed on estimating m and k.

Ehrenberg[2] showed that the number of purchase occasions of a population could be adequately modeled using the Gamma-Poisson process. As an alternative parametrization for the NBD, Ehrenberg used the penetration, $b = 1 - p_0$, and the purchase frequency, $w = \mathbb{E}(X | X \geqslant 1)$. In this paper an alternative parametrization denoted by (b, w') with $w' = 1/w$ will be considered. Its appeal lies in the fact that the corresponding parameter space is within the unit square $(b, w') \in [0, 1]^2$, which makes it easier to make a visual comparison of the behavior of different estimators for all parameter values of the NBD. Fig. 1 shows the contour levels of a, m and k within the (b, w')-parameter space. The NBD is only defined for the parameter pairs $(b, w') \in (0, 1) \times (0, 1)$ such that $w' < -b/\log(1-b)$.

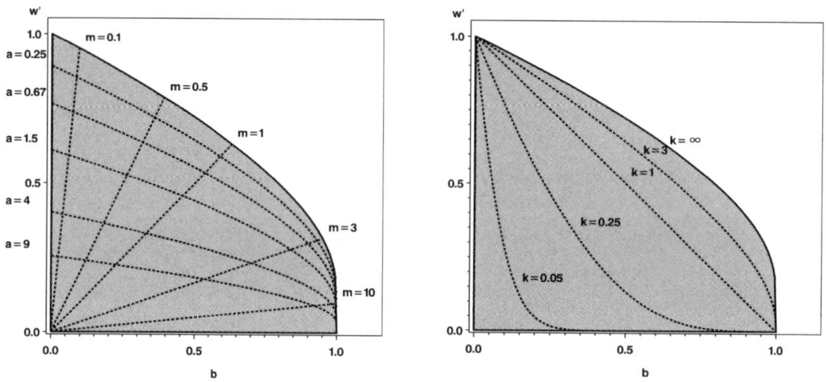

Figure 1. a, m and k versus (b, w').

3. Parameter Estimation for a Unit Time Period

Maximum Likelihood

The ML estimation for NBD parameters was considered independently by Fisher[5] and Haldane.[6] The ML estimator for k cannot be presented in an explicit form and must be found as the solution to an equation. Anscombe[7] noted that the ML equation for k can be tedious to solve and therefore considered moment based estimators for the parameters of an i.i.d. NBD sample. The ML estimator for m is simply given by the sample mean

$$\hat{m} = \bar{x} = \frac{1}{N} \sum_{i=1}^{N} x_i \ ,$$

however there is no closed form solution for $\hat{k}_{ML}$ (the ML estimator of k) and the estimator $\hat{k}_{ML}$ is defined as the solution, in z, to the equation

$$\log\left(1 + \frac{\bar{x}}{z}\right) = \sum_{i=1}^{\infty} \frac{n_i}{N} \sum_{j=0}^{i-1} \frac{1}{z+j} \ .$$

Here N denotes the sample size and n_i denotes the observed frequency of $i = 0, 1, 2, \ldots$ within the sample.

The variances of the ML estimators are the minimum possible asymptotic ($N \to \infty$) variances attainable in the class of all asymptotically normal estimators and therefore provide a lower bound for the moment based estimators. The asymptotic variances of $\hat{m}$ and $\hat{k}_{ML}$ [5,6] are

$$\lim_{N\to\infty} N \operatorname{Var}(\hat{m}) = ka(1+a), \tag{2}$$

$$v_{ML} = \lim_{N\to\infty} N \operatorname{Var}\left(\hat{k}_{ML}\right) = \frac{2k(k+1)(a+1)^2}{a^2\left(1+2\sum_{j=2}^{\infty}\left(\frac{a}{a+1}\right)^{j-1}\frac{j!\Gamma(k+2)}{(j+1)\Gamma(k+j+1)}\right)}. \tag{3}$$

Using the inverse of the Fisher information matrix, the asymptotic covariance between the estimators is $\lim_{N\to\infty} N \operatorname{Cov}(\hat{m}, \hat{k}_{ML}) = 0$ and hence the ML estimators are asymptotically uncorrelated.

Moment Based Estimators

The estimation of the parameter pair (m, k) requires the choice of two sample moments. The first sample moment $\bar{x}$ is both an efficient and an unbiased estimator for the parameter m. An additional moment is then required to estimate the shape parameter k. Denote this moment by $\bar{f} = \frac{1}{N}\sum_{i=1}^{N} f(x_i)$. Anscombe[7] considered various functions $\bar{f}_j = \frac{1}{N}\sum_{i=1}^{N} f_j(x_i)$ for the estimation of k. Let $I_{[x=0]}$ denote the indicator function such that $I_{[x=0]} = 1$ if $x = 0$ and $I_{[x=0]} = 0$ otherwise. Table 1 shows the moment based estimators for k, denoted by $\hat{k}$, for different functions $f_j(x)$.

Note that the PM estimator depends on an additional parameter c ($c > 0$, $(c \neq 1)$). Although an explicit formula exists for the standard methods of moments estimator ($\hat{k}_{MOM}$), no analytical solution exists for the zero term method estimator ($\hat{k}_{ZTM}$), the factorial method estimator ($\hat{k}_{FFM}$) or the power method estimator ($\hat{k}_{PM}$) for k. Since there is at most one solution for $\hat{k}_{ZTM}$, $\hat{k}_{FFM}$ and $\hat{k}_{PM}$, these estimators may be obtained by using numerical algorithms to solve the corresponding equations given in Table 1 for z.

The asymptotic normalized covariance between moment based estimators $\hat{m}$ and $\hat{k}$ is $\lim_{N\to\infty} N \operatorname{Cov}(\bar{x}, \hat{k}) = 0$.[7] Since, amongst the class of moment based estimators considered, the estimator for m is the same and the asymptotic covariances between the estimators of k and m is zero, the most efficient estimation method is determined by the method that minimizes the variance of $\hat{k}$. The asymptotic normalized variances of $\hat{k}_{MOM}$,

Table 1. Moment based estimators for the NBD parameter k.

Method	$f_j(x)$	$\hat{k}$	Estimator or equation for $\hat{k}$
MOM	$f_1(x) = x^2$	$\hat{k}_{MOM}$	$\frac{\bar{x}^2}{\overline{x^2} - \bar{x}^2 - \bar{x}}$
ZTM	$f_2(x) = I_{[x=0]}$	$\hat{k}_{ZTM}$	$\frac{1}{N}\sum_{i=1}^{N} I_{[x_i=0]} = (1 + \frac{\bar{x}}{z})^{-z}$
FFM	$f_3(x) = \frac{1}{x+1}$	$\hat{k}_{FFM}$	$\frac{1}{N}\sum_{i=1}^{N} \frac{1}{x_i+1} = \frac{z}{\bar{x}(z-1)}\left[1 - \left(\frac{z}{\bar{x}+z}\right)^{z-1}\right]$
PM	$f_4(x) = c^x$ $(c > 0,\ (c \neq 1))$	$\hat{k}_{PM}$	$\frac{1}{N}\sum_{i=1}^{N} c^{x_i} = \left(1 + \frac{\bar{x}(1-c)}{z}\right)^{-z}$

$\hat{k}_{ZTM}$ and $\hat{k}_{PM}$ [7,8] are

$$v_{MOM} = \lim_{N\to\infty} N\mathrm{Var}\left(\hat{k}_{MOM}\right) = \frac{2k(k+1)(a+1)^2}{a^2},$$

$$v_{ZTM} = \lim_{N\to\infty} N\mathrm{Var}\left(\hat{k}_{ZTM}\right) = \frac{(a+1)^{k+2} - (a+1)^2 - ka(a+1)}{[(a+1)\log(a+1) - a]^2},$$

$$v_{PM}(c) = \lim_{N\to\infty} N\mathrm{Var}\left(\hat{k}_{PM}\right) = \frac{\left(1+a-ac^2\right)^{-k} r^{2k+2} - r^2 - ka(a+1)(1-c)^2}{[r\log(r) - r + 1]^2}, \quad (4)$$

where $r = 1 + a - ac$. The variance of $\hat{k}_{FFM}$ is difficult to express explicitly and for an expression of the variance we refer to [7, p. 369].

The Power Method of Estimation

The power method of estimation for fixed time intervals has been considered.[8,9] It is proven[9] that the PM estimator, when correctly implemented, is always more efficient than both the MOM and ZTM estimators. Denote the power method estimator for k computed at c as the PM(c) estimator. Let c_o denote the value of c that minimizes $v_{PM}(c)$ for fixed a and k. Figure 2(a) shows levels of c_o within the NBD parameter space and Fig. 2(b) shows the asymptotic normalized efficiency of the PM(c_o) estimator relative to the ML estimator. It is clear that the PM(c_o) estimator is almost as efficient as the ML estimator for the majority of the NBD parameter space.

4. Parameter Estimation for a General Time Period

When considering the efficient estimation of Gamma-Poisson parameters there is the added flexibility of being able to choose the time interval over

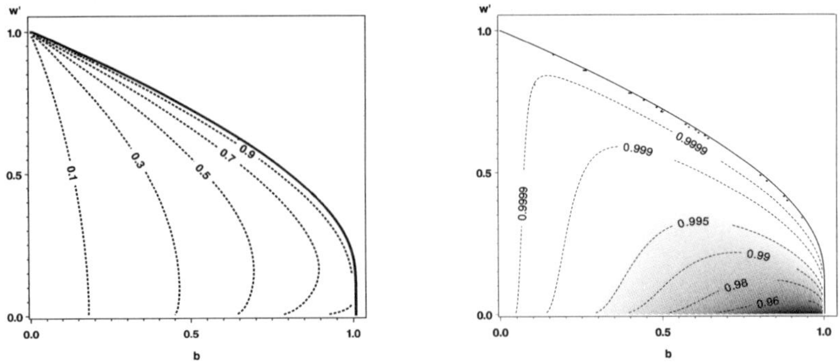

(a) (b)

Figure 2. (a) Contour levels of c_o and (b) Contour levels of $v_{ML}/v_{PM}(c_o)$.

which to collect data. The parameter m varies linearly with time. If m is the mean of the NBD over a unit time interval, the mean of the NBD over a general time interval of length t is mt (follows directly from (1)). The problem, therefore, is to efficiently estimate the parameters (m, k) from a NBD with parameters (mt, k), where t is arbitrary. The parameter m is efficiently estimated by $\hat{m} = \bar{x}_t/t = \sum_{i=1}^{N} x_{i,t}/(Nt)$. The parameter k may be estimated using the estimators shown in Table 1 with $\bar{x}$ replaced by $\bar{x}_t = \frac{1}{N}\sum_{i=1}^{N} x_{i,t}$. The criteria of efficiency is to minimize the variance of the estimators of m and k.

4.1. *Estimating m*

Since the sample mean is an unbiased and efficient estimator for the mean parameter of the NBD, the parameter m is efficiently estimated by $\hat{m} = \bar{x}_t/t = \sum_{i=1}^{N} x_{i,t}/(Nt)$. The variance of this estimator is

$$\lim_{N\to\infty} N \operatorname{Var}(\hat{m}) = \frac{1}{t^2}\operatorname{Var}(\bar{x}_t) = \frac{1}{t^2}kat(1+at) = ka^2 + \frac{ka}{t}$$

where $a = m/k$. The variance for $\hat{m} = \bar{x}_t/t$ is a strictly decreasing function in t and therefore to minimize the variance of $\hat{m}$ the largest value of t possible should be taken.

4.2. *Estimating k Using Maximum Likelihood*

The variance of the maximum likelihood estimator for k is

$$v_{ML}(t) = \lim_{N\to\infty} N \operatorname{Var}\left(\hat{k}_{ML}\right) = \frac{2k(k+1)(at+1)^2}{(at)^2\left(1+2\sum_{j=2}^{\infty}\left(\frac{at}{at+1}\right)^{j-1}\frac{j!\Gamma(k+2)}{(j+1)\Gamma(k+j+1)}\right)}.$$

Consider the derivative of $v_{ML}(t)$ with respect to t for fixed a and k,

$$v'_{ML}(t) = \frac{\partial v_{ML}(t)}{\partial t} = 2k(k+1)\left[-2\frac{at+1}{a^2t^3}Q(t) + \left(\frac{at+1}{at}\right)^2 Q'(t)\right]$$

where

$$Q(t) = \frac{1}{1 + 2\sum_{j=2}^{\infty}\left(\frac{at}{at+1}\right)^{j-1}\frac{j!\Gamma(k+2)}{(j+1)\Gamma(k+j+1)}}.$$

Note that $Q(t) > 0$ and $Q'(t) < 0$ for any $a > 0, k > 0$ and $t > 0$; this implies that $v'_{ML}(t) < 0$ for any $a > 0, k > 0$ and $t > 0$. To minimize the variance $v_{ML}(t)$, it is therefore necessary to take t as large as possible.

4.3. *Estimating k Using Moment Based Estimators*

The variances for the method of moments, power method and zero term method estimators of k are

$$v_{MOM}(t) = \lim_{N\to\infty} N\text{Var}\left(\hat{k}_{MOM}\right) = \frac{2k(k+1)(at+1)^2}{(at)^2},$$

$$v_{ZTM}(t) = \lim_{N\to\infty} N\text{Var}\left(\hat{k}_{ZTM}\right) = \frac{(at+1)^{k+2} - (at+1)^2 - kat(at+1)}{[(at+1)\log(at+1) - at]^2},$$

$$v_{PM}(c;t) = \lim_{N\to\infty} N\text{Var}\left(\hat{k}_{PM(c)}\right) = \frac{\left(1+at-atc^2\right)^{-k} r^{2k+2} - r^2 - kat(at+1)(1-c)^2}{[r\log(r) - r + 1]^2},$$

where $r = 1 + at - atc$.

It is straightforward to check that $v'_{MOM}(t) < 0$ for all $a > 0, k > 0$ and $t > 0$. To minimize the variance of the MOM estimator for k, the largest value possible for t must therefore be taken. Investigating the efficiencies of the ZTM and PM estimators for k prove to be more difficult due to the complex form of the equations for the normalized variances. Figure 3 shows the asymptotic normalized variance of estimators for k using the MOM, PM(0.5), ZTM and ML estimators for two different parameter values of (m, k). Both figures show that, for fixed m and different values of k, there exists optimum values of t and c when estimating k using the PM(c) estimator. Note that the PM(0) estimator is the ZTM estimator.

Figure 4 shows optimum values of c, denoted by c_o, and Fig. 5 shows optimum values of t, denoted by t_o, that minimize $v_{PM}(c,t)$ in the case when the value of t is bounded and $c \in (0, 0.999]$. The value c is bounded for simplicity and practicality, since in cases where $c_o > 0.999$ the function $v_{PM}(c,t)$ is very sensitive to small changes in c and t. In Fig. 5, for $t \in$

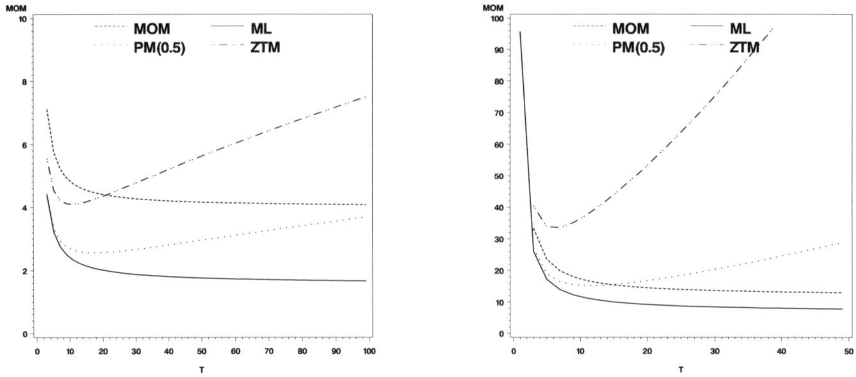

$m = 1,\ k = 1$ $\qquad$ $m = 1,\ k = 2$

Figure 3. $v_{MOM}(t)$, $v_{ZTM}(t)$, $v_{PM}(0.5, t)$ and $v_{ML}(t)$ versus t.

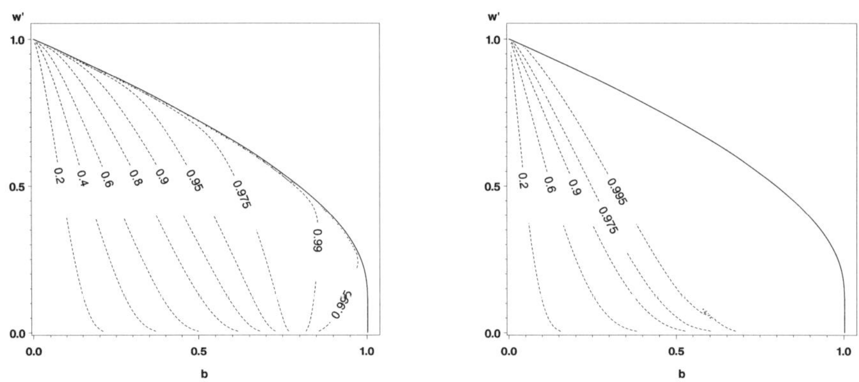

(a) Optimum c for $t \in (0, 100]$ $\qquad$ (b) Optimum c for $t \in (0, 10000]$

Figure 4. Contour levels of c_o in the minimization of $v_{PM}(c, t)$ when $c \in (0, 0.999]$.

$(0, 100]$, the value $t_o = 100$ for the majority of the parameter space. Figure 6 shows the efficiency of $v_{PM}(c_o, t_o)$ for each of the bounded ranges of $t \in (0, 100]$ and $t \in (0, 10000]$ relative to the ML estimator, which is computed at the largest possible value of t within the bound. Taking the largest value of t ensures that the most efficient ML estimator is chosen.

Note that the efficiency in the case $t \in (0, 10000]$ is worse than the efficiency in the case $t \in (0, 100]$. This is because, as t increases, the variance of the estimator for k decreases at a faster rate for the ML estimator in comparison to the PM(c_o) estimator computed at t_o. Note that although the

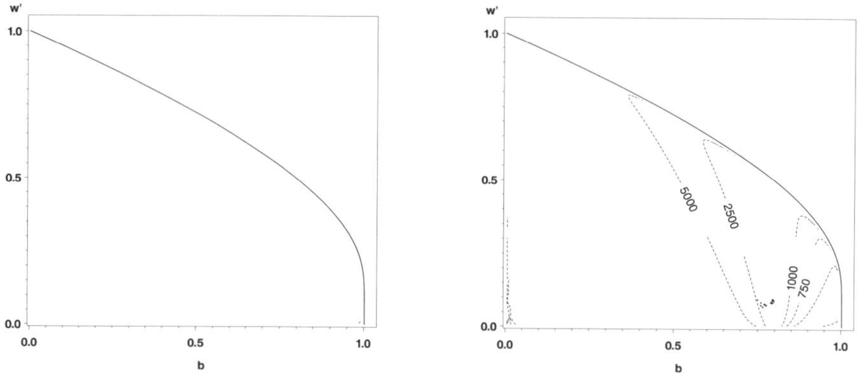

Optimum t for $t \in (0, 100]$ Optimum t for $t \in (0, 10000]$

Figure 5. Contour levels of t_o in the minimization of $v_{PM}(c, t)$ when $c \in (0, 0.999]$.

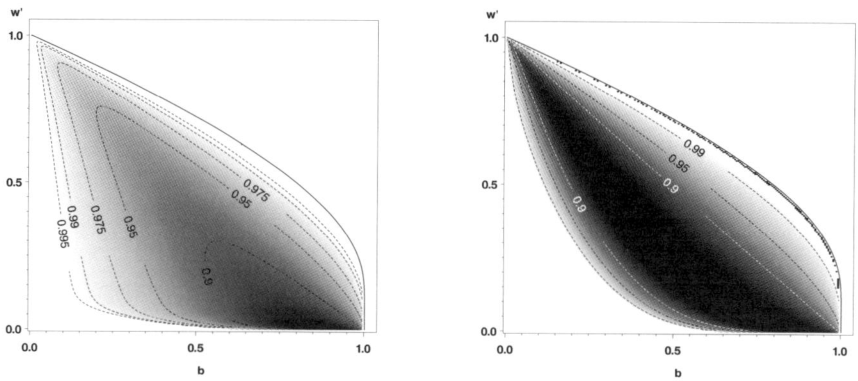

$t \in (0, 100]$ $t \in (0, 10000]$

Figure 6. Efficiency $v_{ML}(t_{ML})/v_{PM}(c_o, t_o)$ where c_o and t_o are values of c and t that minimize $v_{PM}(c, t)$ in the case when the value of t is bounded and and $c \in (0, 0.999]$. The value $t_{ML} = 100$ for $t \in (0, 100]$ and $t_{ML} = 10000$ for $t \in (0, 10000]$.

efficiency of the PM estimator may decrease relative to the ML estimator, it is still possible for the variance of the PM estimator to decrease.

5. Conclusion

The aim of this paper was to investigate the efficient estimation of Gamma-Poisson process parameters. Efficient estimation requires the choice of an optimal time window within which to collect data in order to obtain efficient moment based estimators for the NBD parameters. The efficiency of these moment based estimators is considered to be relative to the maximum likelihood estimators. Maximum likelihood estimators, although efficient in the class of asymptotically normal estimators, are often difficult to implement in practice. If maximum likelihood estimators can be implemented then, since $v_{ML}(t)$ decreases as t increases, a large a time window as possible should be taken to obtain estimators with the smallest possible variance.

For the method of moments estimators since $v_{MOM}(t)$ decreases as t increases a large a window as possible should be taken to obtain efficient estimators for the NBD parameters. For the zero term method estimators and power method estimators, there exists an optimal time t, with $0 < t < \infty$, that minimizes the variance of the estimator for k. This however contradicts to the time interval required to minimize the variance of the estimator of m, where t should be taken to be as large as possible.

For all NBD parameter values and fixed time t, the efficiency of the method of moments and zero term method estimators can be improved by using the power method with $c \in (0, 1)$. If time t is unconstrained then the optimal parameter for c tends very close to 1, although the value of $c = 1$ itself is dismissed as an optimum value.

References

1. O. Lundberg, *On Random Processes and their Application to Sickness and Accident Statistics*, Almquist and Wiksells, Uppsala (1964).
2. A. S. C. Ehrenberg, *Repeat-buying: Facts, Theory and Applications*, Charles Griffin & Company Ltd., London (1988).
3. J. Grandell, *Mixed Poisson Processes* (Vol. 77), Chapman & Hall, London (1997).
4. R. J. Cook and W. Wei, Conditional analysis of mixed poisson processes with baseline counts: implications for trial design and analysis, *Biostatistics* **4**, 479–494 (2003).
5. R. A. Fisher, The negative binomial distribution, *Ann. Eugenics* **11**, 182–187 (1941).
6. J. B. S. Haldane, The fitting of binomial distributions, *Ann. Eugenics* **11**, 179–181 (1941).
7. F. J. Anscombe, Sampling theory of the negative binomial and logarithmic series distributions, *Biometrika* **37**, 358–382 (1950).
8. V. Savani and A. Zhigljavsky, Efficient estimation of parameters of the negative binomial distribution, *Communications in Statistics: Theory and Methods* **35(5)** (2006).
9. V. Savani and A. Zhigljavsky, Efficient parameter estimation for independent and INAR(1) negative binomial samples, *Metrika*, accepted (2006).

A REDUNDANCY DETECTION APPROACH TO MINING BIOINFORMATICS DATA*

H. CAMACHO AND A. SALHI

University of Essex, Wivenhoe Park
Colchester CO4 3SQ, United Kingdom, E-mail:{jhcama, as}@essex.ac.uk

This paper is concerned with the search for sequences of DNA bases via the solution of the key equivalence problem. The approach is related to the hardening of soft databases method due to Cohen *et al.*[1] Here, the problem is described in graph theoretic terms. An appropriate optimization model is drawn and solved indirectly. This approach is shown to be effective. Computational results on bioinformatics databases are included.

1. Introduction

The search for sequences of bases corresponding to genes in the genome has become a crucial problem of medicine and bioinformatics. Genome data is still fresh and yet to be exploited fully. There is a lot of hope to devise new treatments for illnesses such as cancer based on information gleaned from these data. However, the datasets are enormous and searching them, almost for any purpose, is computationally intensive.

In natural language processing, the problem of detecting redundancy in large databases has been considered for many years. Although not yet satisfactorily solved due to its inherent complexity, many useful methods have been devised for it. These approaches may be different, but all of them measure in one way or another the similarity between records containing symbols of the alphanumeric type. Accuracy and computational efficiency is what separates them. Unlike bioinformatics, a lot of these techniques are mature.

Since genome data is text-based (symbols of the alphabet) approaches such as record linkage,[2,3] hardening,[1] merge/purge,[4] and record matching[5] must, in principle, be applicable. However, the bioinformatics problem

*This work is supported by CONACYT grant 168588.

must be cast in an appropriate form.

The case of interest concerns the scanning of genome data for probes, such as the Affymetrix 25-base probes,[6] which are used to measure mRNA abundance.[7] Approaches which consider the similarity of chemical components also exist.[8]

Initially, the genome data, or a subsequence of it, is sliced into sequences of bases (C, G, T, A) of a certain length to match that of the probes. These sequences are nothing more than strings or words of the alphabet $\{C, G, T, A\}$. Each one is then stored as a record in a database containing the probe(s) and then the task of searching for redundancy of records in this database can be approached as the key equivalence problem. The latter has been investigated recently through the *Hardening of Soft Information Sources* approach of Cohen *et al.*,[1] requiring the solution of a global optimization problem.

Our approach to the problem is related, but simpler.[9] Although it is also formulated as an optimization problem, the latter is more tractable than global optimization. This simplification follows from the fact that a record has potentially many fields each pointing to a real world object, i.e. it forms a reference. Here, we consider that the whole record, however many fields it may have, points to a single object. This is an important distinction since the initial complete graph we work from is less complex than what would be considered if the model used was exactly adhered to.

The present work explains how this can be done and reports results on real data from Affymetrix.[a6] In section 2 the key equivalence problem is formulated, in section 3 the solution approach is defined. In section 4 experimental results are reported. Section 5 is the conclusion.

2. Formulation of the Key Equivalence Problem

Let *object identifier* O_i be any record in a database corresponding to each of the 25-base probes sequences. Let also *object* be the real target to which O_i is referring, and *key* the unique identification of the record in a database. Then, *key equivalence* occurs when two or more O_i's in a database refer to the same *object*.[10] As said earlier, the main difference between our formulation and that of the hardening approach,[1] is that here we consider a database as a set of O_i's, while in Cohen *et al.*'s work, a database consists

[a] Affymetrix is a divsion of Affymax, a bioinformatics company formed in 1991. It is dedicated to developing state-of-the-art technology for acquiring, analyzing, and managing complex genetic information for use in biomedical research.

of a set of *tuples*, each of which consisting of a set of *references*, or fields. Each reference points to a real world object.

Since, given a database, it is not easy to tell which records point to the same object, we initially assume that all of them point to the same object. This means that all records can potentially be represented by the same object identifier. Therefore, initially at least, we in fact assume that when all redundancy is removed, we will possibly be left with no database. This assumption may sound unreasonable, since only a small percentage of records in a database might be corrupted, but it is necessary to motivate our method. Moreover, it does not limit the application of the method suggested.

Let now each object identifier be represented by a node. Then, the *potential* redundancy of an identifier may be represented by a directed arc between this identifier and another one. An incoming arc means the source node is potentially redundant. Since, as was assumed, initially they all point to each other, no direction is required, leading to a complete graph.

Let $G(V, E)$ be this graph with $V = \{1, 2, ..., i, ..., n\}$ its set of nodes, each corresponding to an object identifier, and $E = \{(i, j)|i, j = 1, 2, ..., n, i \neq j\}$ its set of arcs. By some string similarity metric, it is possible to find weights for all edges of graph G specifying how likely it is that two object identifiers point to the same real world object, i.e. one of them is redundant. A large weight between two O_i's means they are unlikely to point to the same object, and a small weight means otherwise, i.e. there is redundancy. In this fashion, since a given normalized string similarity takes values in $[0, 1]$, where 1 is the maximum similarity, we take as a weight its inverse value (1$-$string similarity). We are, now, left with the question of how close to zero a weight has to be in order to say that one of the records is redundant. Clearly, a subgraph of G with minimum total weight will catch redundancy. Moreover, this subgraph must have all the nodes of G.

3. Solution Approach

A further formalization is necessary to model this situation. In particular, we consider that a subgraph of G that captures all or part of the redundancy in the database, is generated by a function from V to V. As such, it has the properties of totality and unicity. Given G, we want to find $G'(V, E')$

such that $E' \subseteq E$, and

$$z = \sum_{(i,j)\in E, i\neq j} e_{ij}w_{ij} + \left(n - \sum_{(i,j)\in E, i\neq j} e_{ij}\right)\lambda_1 + \left(\sum_{(i,j)\in E, i\neq j} e_{ij}\right)\lambda_2 \quad (1)$$

is minimized. In z, $e_{ij} = 1$ if $(i,j) \in E'$ and 0, otherwise, n is the size of the database, w_{ij}, i, $j = 1, 2, ..., n$ are the weights, and λ_1 and λ_2 are constants which control the size of the resulting database for the amount of redundancy detected. Equivalently, they are constants which, when exactly known, will give a value z which is smallest for the database that has been cleaned of all its redundancy and nothing else, i.e. the perfect solution. Of course the choice of these constants will influence the effectiveness of the approach advocated here. By constraining z with the requirements of the relation (function) between the nodes, and by a simple manipulation of the expression due to the fact that some terms are constants, by replacing $\lambda_1 - \lambda_2$ with a single parameter k, we obtain the following optimization problem.

$$\min z = \sum_{(i,j)\in E, i\neq j} e_{ij}w_{ij} - k \sum_{(i,j)\in E, i\neq j} e_{ij} \quad (2)$$

$s.t.$

$$\sum_{j\in V} e_{ij} \leq 1, \forall i, i \neq j \quad (3)$$

$$\sum_{(i,j)\in E} e_{ij} \leq n - 1 \quad (4)$$

$$u_i - u_j \leq n - 1 - ne_{ij}, \quad (5)$$
$$i, j = 1, ..., n, i \neq j, u_i \in R_+$$

Note that restrictions (3) imply that there is at most one edge (i,j) from each node i. Restrictions (4) and (5) eliminate cycles.[11]

From the above model, it is clear that if $k \leq 0$, the second term of z is zero or positive and so the minimum corresponds to all $e_{ij} = 0$, i.e. no edge is worth including in the solution, giving $E' = \emptyset$. If $k > 0$ the minimum of z must be negative, i.e. $\sum_{(i,j)\in E, i\neq j} e_{ij}w_{ij} \leq k \sum_{(i,j)\in E, i\neq j} e_{ij}$ in which case the solution to the above model will be those arcs with small weights. Moreover, because we are minimizing, all these weights will be less than k.

3.1. *Formalization*

Parameter k is essential for trapping redundancy and its proper setting will decide on how successful the detection of redundancy will be. Set too large, a connected subgraph of G will be the solution, thus including all nodes (object identifiers). Set too low, very few, if any, will be included in the solution, thus leaving out genuine redundancy. It must be clear already that trees satisfy the constraints of the above optimization model. A solution to the problem is likely to be a collection of subtrees of the minimum spanning tree of G. In other words, it is most likely to be a forest.

Definition 3.1. A spanning forest of a connected graph G is a forest whose components are subtrees of a spanning tree of G.

Definition 3.2. A minimum spanning forest of a connected graph G is a forest whose components are minimum spanning trees of the corresponding components in G.

Proposition 1 *The solution to the suggested optimization model is a spanning forest (tree). Moreover, it is a minimum spanning forest (tree).*

Proof. **Case 1**: The database is totally redundant, i.e. it can be represented by a single record.

Let G be the complete graph of the database in hand, and assume that the minimum spanning tree of G is found. Because all nodes are similar to each other, $\forall(i,j)$ in the minimum spanning tree of G, $w_{ij} \leq k$, k being the similarity threshold chosen as the weight of the arc linking the two similar records (nodes) with largest weight. Therefore redundancy is clearly captured by a minimum spanning tree.

Case 2: There are at least two records which are not similar to each other.

In this case, the minimum spanning tree of G will have at least two arcs (i,j) such that $w_{ij} > k$, k being the similarity threshold chosen as the weight of the arc linking the two similar records (nodes) with largest weight. These arcs will be removed from the minimum spanning tree since the linked records are different, according to the chosen k. We know that removing an arc from a tree always results into a disconnected tree, or forest, the redundancy in this case must be captured by a minimum spanning forest.

Case 3: All records are different from each other, i.e. there is no redundancy. In this case k must be 0, which means that all edges of the minimum spanning tree of G are removed, leaving a minimum spanning forest which is completely disconnected. □

Proposition 2 *For a given k, the optimal solution to model (2)-(5) can be obtained in polynomial time.*

Proof. Apply a greedy algorithm to the complete graph of G of the database, to find its minimum spanning tree. Choose an appropriate similarity threshold k. Trim the graph of all its edges with weights greater that k. The remaining subgraph is either a tree or a forest. Both steps can be done in polynomial time. □

3.2. *Estimating k*

The threshold constant k can be chosen arbitrarily, below 0.5, for instance. That may well work in some cases. However, in general, it is better to find an estimate directly related to the given database, (see *Remark 2* below). This can be done as follows.

3.2.1. *Algorithm 1:*

(1) Find the weighted complete graph G corresponding to the given database;
(2) Find the minimum spanning tree of G;
(3) Assign to k the largest weight for which a good match between records is found;

Remark 1 *The similarity threshold k used is only an approximate value.*
Remark 2 *The problem of finding the optimum k may not be solvable in polynomial time. This is because the concept of "a good match" cannot be be defined accurately. Also, the length of records may require exponential-time computing procedures to decide whether there is any reasonable similarity. Algorithm 1 is only a practical approach to estimating k.*
Remark 3 *Parameter k varies from database to database.*

3.3. *Detecting Redundancy*

Redundant records are detected according to the following algorithm.

3.3.1. *Algorithm 2:*

(1) Apply Algorithm 1 to the given database;

(2) Remove all edges with weights $> k$, from the minimum spanning tree of G output by Algorithm 1.

The output is a tree (or forest) that represents the detected redundant records. Each tree of the spanning forest can be reduced to one node. The remaining nodes of the forest constitute the records of the resulting database after removing redundancy.

4. Experimental Results

4.1. *Weight Computation*

Because of the special structure of the database generated from bioinformatics data, we suspected that some similarity metric may be more suitable in terms of accuracy than others. So, to compute the required weights, a number of string metrics are investigated, and the best performing string metric according to standard performance measures, is retained as the weight provider for our method. Below is the outcome of such an investigation.

We generated 5 artificial datasets containing rows of 25-character long sequences with different sets of corrupted redundant records with different probability values, ranging from 0.1 to 0.5. We implemented the string metrics contained in the SecondString[12] and Simmetrics packages.[13]

To evaluate the string metrics, we compute the non-interpolated average precision of the ranking based on the computed weights using different metrics, as well as the harmonic mean at position l given by the maximum of $F1(l), l = 1, ..., N$, with $F1(l) = \frac{2*precision*recall}{precision+recall}$, as in Cohen *et al.*[12] There, a third measure, the interpolated precision of recall was also calculated, which is not here.

The evaluations are carried out as follows. Given a database, let $N = |E|$ be the number of candidate pairs ranked by score. Let m be the number of actual matching pairs, or redundancies in the database. We calculate the non-interpolated average precision $\tau(c,d) = \frac{1}{m}\sum_{l=1}^{N}\frac{c(l)d(l)}{l}$, where $c(l)$ is the number of true matching pairs before rank position l, $d(l)$ a binary indicator, which takes value 1 if the lth pair is a true match, and value 0 otherwise.[12] We also calculate $max_{l>0}F1(l)$, the harmonic mean at rank position l, $precision = \frac{c(l)}{l}$, and $recall = \frac{c(l)}{m}$.

The best performing methods out of about 30 are shown in Table 1. Among these, the Needleman-Wunsch method performed best on average.

Table 1. String metric comparison over artificial datasets.

Source	Method	$\tau(c,d)$	$max_{l>0} F1(l)$
1	NeedlemanWunch	0.760	0.774
1	Levenshtein	0.680	0.694
2	Level2MongeElkan	0.594	0.628
2	MongeElkan	0.594	0.628
1	SmithWatermanGotoh	0.578	0.604
1	MongeElkan2	0.578	0.604
1	QGramsDistance	0.372	0.406
1	JaroWinkler	0.359	0.407
2	Jaro	0.266	0.337

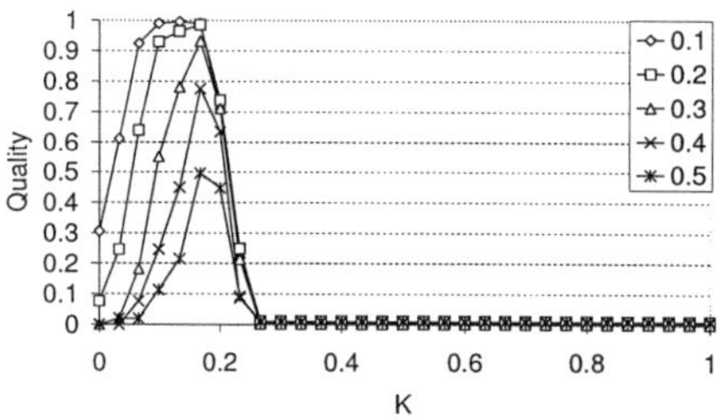

Figure 1. Quality of performance in 5 artificially generated datasets. The level of corruption is ranged from 0.1 to 0.5.

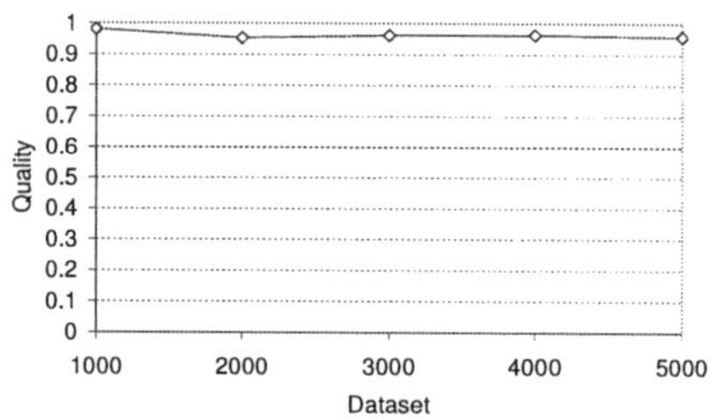

Figure 2. Performance evaluation of proposed method over five artificially generated datasets. The level of corruption ranges from 0.1 to 0.5.

4.2. *Test Dataset Results*

We implemented Algorithms 1 and 2 and the Needleman-Wunsch metric mentioned earlier to provide the weights. These implementations were then tested on the 5 artificial datasets. Here, the $F1$ factor is evaluated at 31 different values of k, between 0 and 1. The results are displayed in Fig. 1 which shows that the optimal value of k ($k = 0.2$) is the same for all the datasets.

4.3. *Larger Datasets*

In order to test the robustness of the proposed method for larger examples, we artificially generated 5 datasets, of 1000, 2000, ..., 5,000 records, each of them with the same probability of corruption (0.2). We evaluated those datasets in the same way as previously mentioned. The results are shown in Fig. 2. Notice that the performance of the proposed method is not sensitive to the size of the datasets.

Table 2. Redundant records detected in real data.

Record T	Record U	Weight
CCCTAACCCTAACCCTAACCCTAAC	CCCTAACCCTAACCCTAACCCTAAC	0
TAACCCTAACCCTAACCCTAACCCT	TAACCCTAACCCTAACCCTAACCCT	0
AGAAAGAAAGAAAGAAAGAAAGAAA	AGAAAGAAAGAAAGAAAGAAAGAAA	0
GTGTGTGTGTGTGTGTGTGTGTGTG	GTGTGTGTGTGTGTGTGTGTGTGTG	0
TAACCCTAACCCTAACCCTAACCCT	TAACCCTAACCCTAACCCTAACCCT	0
TAACCCTAACCCTAACCCTAACCCC	TAACCCTAACCCTAACCCTAACCCT	0.02
AACCTAACCCTAACCCTAACCCTAA	ACCCTAACCCTAACCCTAACCCTAA	0.02
AACCCTAACCCTAACCCCTAACCCT	AACCCTAACCCTAACCCCTAACCCC	0.02

4.4. *Real Dataset*

We used part of the human genome as found in the ensembl project[14] literature. We split this sequence into 25-base long subsequences. With these subsequences, we compiled a dataset with 10,000 records. We set the similarity threshold value k to 0.2. This value was pointed out by the results of Fig. 1. A list of detected redundant records is shown in Table 2.

5. Conclusion

We have looked at the problem of detecting and removing redundancy in datasets, and we suggested a solution approach. In particular, we solved the problem of detecting similar records in 25-base long sequences from the human genome data. This is an important up-to-date problem in bioinformatics.[15] An optimization model of the integer programming type has been devised for it. Although, this model is difficult to solve directly, it turns out that a tree graph of a certain kind (a forest) provides an optimum solution, when the all important similarity threshold parameter k, was provided. Moreover, it was shown that in this case, the solution can be obtained in polynomial time.

Because the weights measuring the similarity between records are of paramount importance, we also looked at existing methods (or string metrics) for computing them and evaluated them over a set of artificially generated datasets. This allowed us to use the best weights for the method suggested.

The reported experimental results show that the proposed method (Algorithm 1 and Algorithm 2) together with the selected string metric is performing well in terms of the quality of the detected matches. Also, because datasets, particularly in bioinformatics, are often very large, we investigated how sensitive the method is to much larger sets than the ones considered here, (Fig. 1). The results show that the suggested method is not sensitive to size; the quality of the solution obtained in all cases remained high, (Fig. 2). However, one should note that in these experiments the levels of corruption were set by hand. Therefore, it is fairly easy to measure how effective the method is. In real applications, there is no knowledge of the level of corruption and redundancy. Measuring the performance of a given method is therefore more difficult. For this reason, such an experiment was also conducted and the results are reported in Table 2.

References

1. H. Kautz, W. Cohen and D. McAllester, Hardening soft information sources, In: *KDD: Proceedings of the international conference on Knowledge discovery and data mining*, Boston, Massachusetts, USA (2000).
2. A. P. James, H. B. Newcombe and S. J. Axford, *Science* **130**, 954–959 (1959).
3. I. P. Fellegi and A. B. Sunter, *Journal of the American Statistical Society* 64 number **328**, 1183–1210 (1969).
4. M. A. Hernandez and S. J. Stolfo, The merge/purge problem for large databases, In: *SIGMOD: Proceedings of the International conference on Management of data* (1995).
5. A. E. Monge and C. P. Elkan, An efficient domain-independent algorithm for detecting approximately duplicate database records, In: *SIGMOD: Proceedings of the workshop on data mining and knowledge discovery* (1997).
6. Affymetrix genechip technology, In: *http://www.affymetrix.com*, Santa Clara, CA (2006).
7. D. Greenbaum, C. Colangelo, K. Williams and M. Gerstein, *Genome Biology* **4**, 117 (2003).
8. V. Monev, *Communications in Mathematical and in Computer Chemistry* **51**, 7–38 (2004).
9. H. Camacho and A. Salhi, A graph theoretic approach to key equivalence, In: *MICAI 2005: Advances in Artifical Intelligence, proceedings of the 4th Mexican International Conference on Artificial Intelligence, LNAI 3789*, pages 524–533, Monterrey, Mexico (2005).
10. C. Pu, Key equivalence in heterogeneous databases, In: *Proceedings of the international workshop on research issues in data engineering* (1991).
11. L. A. Wolsey and G. L. Nemhauser, *Integer and Combinatorial Optimization*, Wiley-Interscience (1999).
12. W. Cohen, P. Ravikumar and S. E. Fienberg, A comparison of string distance metrics for name-matching tasks. In: *IJCAI and IIWEB*, Acapulco, Mexico (2003).
13. S. Chapman, *Simmetrics Web Intelligence*, Natural Language Processing Group, Department of Computer Science, University of Sheffield, Regent Court, 211 Portobello Street, Sheffield, S1 4DP, UK. sam@dcs.shef.ac.uk (2006).
14. T. Hubbard *et al.*, Ensembl nucleic acids res, In: *33 Database Issue: D447-53* (2005).
15. A. Harrison, Private discussion of bioinformatics data, University of Essex (2005).

OPTIMAL OPEN-LOOP RECIPE GENERATION FOR PARTICLE SIZE DISTRIBUTION CONTROL IN SEMI-BATCH EMULSION POLYMERISATION

N. BIANCO AND C. D. IMMANUEL*

Department of Chemical Engineering, Centre for Process System Engineering, Imperial College London, London SW7 2AZ, UK

Emulsion polymerisation is widely used in industry to produce a large range of products. Many important properties of the polymer products are strongly influenced by the particle size distribution (PSD) of the latex. PSD is driven by three major phenomena: particle nucleation, growth and coagulation, each of which strongly interact with each other, present irreversible traits and have widely different time constants, thereby rendering PSD control challenging. In this study a population balance model is used to develop feed policy to attain target PSD. A multi-objective optimisation strategy that targets the individual phenomena of nucleation, growth and coagulation is adopted.

1. Introduction

Emulsion polymerisation is widely used in large-scale industrial productions. Products obtained by emulsion polymerisation include paints, adhesives, coatings, foams, films, cosmetics and synthetic rubbers. Important characteristics of these products such as mechanical strength, rheological, adhesion and film-forming properties are strongly influenced by the particle size distribution (PSD) of the final latex. Typical application also necessitate attainment of complex and multi-modal PSD.[1] These factors justify the need for control studies on the PSD of the emulsion latex.[2]

In the current research the features of a comprehensive model for emulsion polymerisation processes based on population balance equations[3,4,5] and a novel numerical solution technique[6,7] are exploited to apply different optimisation strategies to the process. These open-loop control studies aim at producing optimal recipes that match target PSD. The copolymerisation of Vinyl Acetate (VAc) and Butyl Acrylate (BuA) under non-ionic

*To who correspondence should be addressed: c.immanuel@imperial.ac.uk

surfactant and redox initiator in a semi-batch reactor is considered.

While a considerable amount of work has been done on the modelling of the process and its numerical solution, there are relatively few control studies. Control of parameters such as PSD[8,9,10] and molecular weight distribution (MWD)[11] have been addressed only in recent years as a direct consequence of the advances in computation techniques and on-line measurement instrumentations.[12]

In controlling the process different approaches can be used. These can be divided in open-loop (feedforward)[13,14] and feedback[8,9] control strategies. The focus of this paper will be on open-loop control scheme, as basis for the development of future closed-loop controls. Online recomputation of open-loop scheme in a feedback configuration have been proved to be effective in producing latexes with defined end-use properties.[15,16]

This paper refers to the work of Immanuel and Doyle III,[10,13] who implemented open-loop control strategies to the copolymerisation of VAc and BuA with the aim to control the full PSD. The optimal recipes were produced following two different strategies. In the first the PSD was controlled directly using single-objective cost functions to optimise the feed rates of surfactant and VAc.[13] These were implemented in a semi-batch reactor to validate the results. In the second a hierarchical approach was used accounting, in this case, for the interaction between the different phenomena that drive the PSD.[10] The optimal recipes were calculated through a multi-objective optimisation strategy, aiming to minimise deviations from the target PSD, as well as target profiles of the number of particles and solids content distribution. The optimisation were carried out employing a Genetic Algorithm, a global optimisation technique.

The main novelties presented in the current paper are the use of gradient-based optimisation strategies to calculate the optimal inputs and the use of a novel solution strategy for population balance models, which quicken the computation of the model compared to other solution techniques.

2. Particle Size Distribution Modelling and Numerical Solution

Emulsion polymerisation is a heterogeneous free radical polymerisation in which polymer particles ($\simeq$ 10 nm - 1 μm) are dispersed in an aqueous phase. The polymer particles are the main locus for the polymerisation reaction. Both continuous and discrete growth take place in the reactor.

Continuous growth is a consequence of the polymer chains growth inside the particles while discrete growth takes place by coagulation of smaller particles to form larger particles. Continuous growth and coagulation, together with nucleation of the particles are the three major phenomena that drive the PSD.

These three phenomena present irreversible traits and strongly interact with each other. A multi-scale nature of the process derive from the different time constants of the particle level processes, as well as due to the multiple level of process details including particle level details and macroscopic details. Any simplifications of the model is precluded by the strongly size-dependent kernels, further adding to control challenge in the form of a competitive growth, for example. To account simultaneously for all these aspects and have a whole picture of the process, as well as to meet complex and multi-modal PSD objectives, a model-based control strategy seems to be the more suitable and so an accurate model is needed.[2]

The model used in the current study account for the copolymerisation of VAc and BuA and was developed for a semi-batch reactor where non-ionic surfactant and redox initiator were employed.[3,4] The model is based on population balance equations. One of them is used to describe the PSD and is expressed for the particle density function $F(r,t)$

$$\frac{\partial}{\partial t}F(r,t) + \frac{\partial}{\partial r}\left[F(r,t)\Re_{growth}(r,t)\right] = \Re_{nuc}(r,t) + \Re_{coag}(r,t) \;, \qquad (1)$$

where r is the radius of the particles, chosen as internal coordinate to describe the distribution of the particle density on time and space. The three major phenomena that drive the PSD are all accounted for in the equation through the rate of nucleation $\Re_{nuc}(r,t)$, growth $\Re_{growth}(r,t)$ and coagulation $\Re_{coag}(r,t)$.

The solution technique used to compute the model is based on a novel solution technique for population balance models named hierarchical two-tier algorithm.[6] As the name suggests the numerical solution is based on two tier. During the first tier the rate of nucleation, growth and coagulation in (1) are calculated and used to update the PSD in the second tier using a formulation that is different from (1). The PSD is updated at each time step until the end of the batch is reached. At each iteration the particle density function $F(r,t)$ is calculated in each of the 250 bins of 2 nm used to discretise the particle size domain.

This solution technique is based on the discretisation of the population into finite elements (or bins) and the formulation of the population balance equation in each bin. Because of this no finite difference discretisation of (1) is needed, thus reducing the truncation-related inaccuracies. The reformulation of the population balance in terms of lumped sub-populations also reduces the stiffness in the solution.

Moreover, the structure of this algorithm constitute a natural framework for the implementation of the so-called multi-level discretisation technique,[7] which introduces a further reduction in the computation time. Exploiting this feature a finer discretisation was used to calculate the nucleation and growth rate, while a coarser discretisation was employed for the computation of the coagulation rate, due to its reduced sensitivity to the width of the bins.

In the present study, the discretisation of the population results in approximately 510 ordinary differential equations, which are solved together with other 30 algebraic equations. These calculations are performed in about 8 s by a computer Pentium IV 2 GHz processor with 512 Mb memory using the multi-level hierarchical two-tier algorithm. Fortran language is used to code the model.

3. Open-loop Control Studies

Open-loop control studies were carried out exploiting the feature of a population balance model and a novel solution technique for population balances to produce optimal recipes that match target PSD. A gradient based optimisation routine based on sequential quadratic programming (SQP) was implemented using subroutines from NAG in FORTRAN.[17]

Sets of experiments were performed using different single objective formulations of cost functions and manipulated variables. The objective functions used were the following

$$\theta_1 = \int_{r_{nuc}}^{r_{max}} \Big(W(r,t_f) - W_{ref}(r)\Big)^2 dr \ , \tag{2}$$

$$\theta_2 = \int_{t=0}^{t_f} \Big(sc(t) - sc_{ref}(t)\Big)^2 dt \ , \tag{3}$$

$$\theta_3 = \int_{t=0}^{t_f} w(t)\Big(Np(t) - Np_{ref}(t)\Big)^2 dt \ , \tag{4}$$

where θ_1 is the 2-norm of the error between end-point weight-averaged PSD and a target weight-averaged PSD. The other two cost functions account for the error on the solids content (sc) and on the number of particles (Np), respectively. The weight $w(t)$ in (4) accounts for the order of magnitude increase in the number of particles, which takes place at each nucleation event.

Sensitivity studies on the copolymerisation of VAc and BuA[18] outlined the effectiveness of surfactant and VAc feed rate as decision variables for the control of the PSD. Exploiting these results a batch spanning 150 minutes was divided in 11 intervals and the feed rates of these reactants were defined as piece-wise constants in each time interval. Hence, the feed rates of surfactant and VAc were used as manipulated variables. In further sets of experiments, the duration of the injection time intervals were also used as manipulated variables in addition to the feed rates of surfactant and VAc.

The first optimisation problem studied was a single-objective formulation aimed at minimising the objective function θ_1:

$$\begin{aligned} &\min_{\mathbf{u}_f} \quad \theta_1 \\ &s.t. \quad \mathbf{a}_f \leq \mathbf{u}_f \leq \mathbf{b}_f \,. \end{aligned} \tag{5}$$

The second optimisation problem was the minimisation of the objective function θ built as the weighted sum of the three objective functions defined above:

$$\begin{aligned} &\min_{\mathbf{u}_f} \quad \theta = \theta_1 + \alpha_2\theta_2 + \alpha_3\theta_3 \\ &s.t. \quad \mathbf{a}_f \leq \mathbf{u}_f \leq \mathbf{b}_f \,. \end{aligned} \tag{6}$$

The third optimisation problem was a min-max problem:

$$\begin{aligned} &\min_{\mathbf{u}_f} \quad \theta = \max(\theta_1, \alpha_2\theta_2, \alpha_3\theta_3) \\ &s.t. \quad \mathbf{a}_f \leq \mathbf{u}_f \leq \mathbf{b}_f \,. \end{aligned} \tag{7}$$

In the above problems, $\mathbf{u}_f \in \mathbf{R}^{n_f}$ is the vector of decision variables representing the first five piece-wise constants feed rates of surfactant and the first six piece-wise constants feed rates of VAc. The terms $\mathbf{a}_f, \mathbf{b}_f \in \mathbf{R}^{n_f}$ are the vectors of simple bounds, lower and upper, used as constraints for the feed rates. n_f is the number of feed rates used as manipulated variables, equal to 11. The weighting of the objective function is used to make the values of the different cost functions comparable.

The distributions attained from the application of these optimisation formulations are showed in Fig. 1. The target was produced through the simulation of the model using specified values for the feed trajectories.

The match between simulated and target distributions is very good. Studying the optimal inputs produced by the optimiser, it is possible to

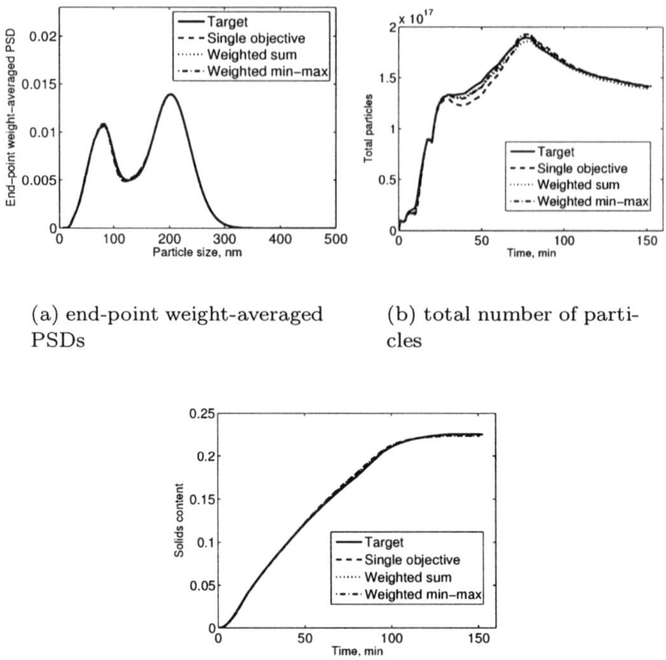

(a) end-point weight-averaged PSDs

(b) total number of particles

(c) solids content

Figure 1. Comparison of weight-averaged PSD, total number of particles and solids content obtained from the three different formulations of the optimisation problem: single-objective, weighted sum and weighted min-max.

notice that the surfactant feed rates follow closely the trajectory used to produce the target while the same is not true for the VAc feed rates. This result could indicate either a larger sensitivity to the surfactant feed rate in the process or the non-convexity of the problem.

A second set of simulations employed the same formulation (5), (6), (7) stated above, but aimed to study whether the increase of the number of decision variables could improve the match with the target distributions. To the surfactant and VAc feed rates were added the durations of their injection times. Three case were analysed, where the first two, first four and first seven time intervals were added to the previous 11 feed rates. Linear inequality constraints were introduced between the time intervals to account for the progression of the batch

$$\begin{aligned} \min_{\mathbf{u}_f,\mathbf{u}_t} \quad & \theta \\ s.t. \quad & \mathbf{a}_f \leq \mathbf{u}_f \leq \mathbf{b}_f \\ & \mathbf{a}_t \leq \mathbf{u}_t \leq \mathbf{b}_t \\ & u_{t,i} \leq u_{t,i+1} \qquad\qquad i = 1, \ldots, n_t - 1 , \end{aligned} \tag{8}$$

where $\mathbf{u}_t \in \mathbf{R}^{n_t}$ is the vector of decision variables representing the batch sub-intervals. The terms $\mathbf{a}_t, \mathbf{b}_t \in \mathbf{R}^{n_t}$ are the vectors of simple bounds, lower and upper, imposed as constraints to them. The parameter n_t is the number of sub-intervals used as manipulated variables.

No improvements in terms of match with the target were achieved using an increased number of variables. Further studies with increased number of variables, eventually using other variables such as feed rates of initiator or water instead of the sub-interval durations, are still motivated due to their larger flexibility and the consequent capability to tackle more complex targets.

4. Pareto Optimisation

To deal with the conflicting/interacting nature of the sub-processes involved in emulsion polymerisation, a multi-objective strategy was employed to calculate optimal recipes. An ϵ-constrained optimisation strategy was employed in this pareto optimisation. Pareto solutions are a set of non-dominated solutions in which each solution is better than the others at least for one of the objective functions.

The objective function θ_1 was minimised subject to non-linear constraints on θ_2 and θ_3:

$$\begin{aligned} \min_{\mathbf{u}_f} \quad & \theta_1 \\ s.t. \quad & \mathbf{a}_f \leq \mathbf{u}_f \leq \mathbf{b}_f \\ & \theta_2 \leq \epsilon_2 \quad ; \quad \theta_3 \leq \epsilon_3 \quad ; \quad \theta_2 \leq \epsilon_2 \ \ and \ \ \theta_3 \leq \epsilon_3 \end{aligned} \tag{9}$$

Three different cases were analysed, two two-objective formulations in which the objective functions θ_2 and θ_3 were subjected to upper bounds respectively and a three-objective case in which both θ_2 and θ_3 were subjected to ϵ-constraints simultaneously. The 11 variable case was considered with only the feed rates of surfactant and VAc used as manipulated variables. Different initial guesses were implemented. The optimal results and

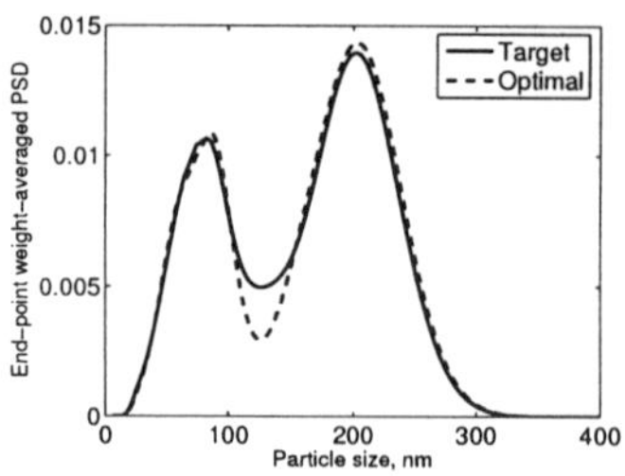

(a) end-point weight-averaged PSDs

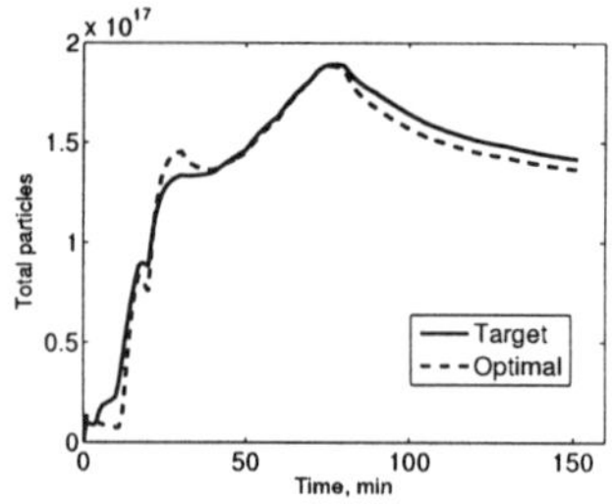

(b) total number of particles

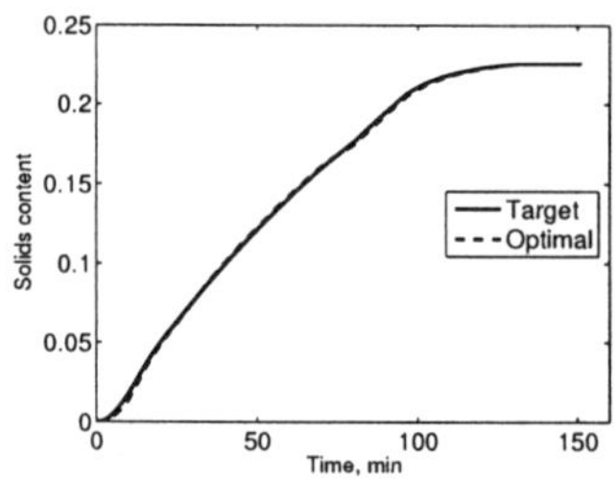

(c) solids content

Figure 2. Comparison between the weight-averaged PSD, total number of particles and solids content for a pareto optimisation based on the three-objective formulation.

final distributions for a three-objective case are shown in Figs. 2 and 3. The original recipe used to produce the target is referred to as "solution" in Fig. 3.

Good matches were observed for all the targets and formulation implemented. A slightly better match was observed using the three-objective formulation confirming the effectiveness of the hierarchical strategy in decoupling the underlying particle rate processes of nucleation, growth and coagulation.[10]

Further pareto optimisation were carried out using a reduced number of decision variables. To verify the possibility of eliminating the VAc feed rates as a manipulated variables, only feed rates of surfactant were used as decision variables.

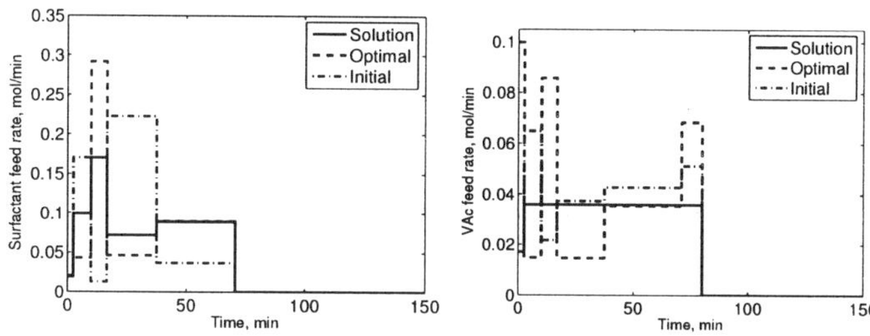

(a) original and optimal feed rates of surfactant

(b) original and optimal feed rates of VAc

Figure 3. Comparison between original and optimal decision variables trajectories for a pareto optimisation based on the three-objective formulation.

While the match with the target PSD was still good, with this reduced set of decision variables the optimiser was unable to ensure a good match with the target profile of total particles and solids content. This confirms the need for multiple actuation in order to bring about independent control of particle level processes of nucleation and growth in addition to the overall PSD.[10] This is a proof of the importance of VAc feed rates as manipulated variables to decouple the nucleation and growth rates. On the other hand, the reduction of the number of variables resulted in a decrease of the optimiser computation load and practically halved the computation time. This result is important in view of online control implementation of the optimiser.

5. Conclusions

In this study, off-line optimisation were performed as a means of open-loop control development. Despite discontinuities in the process and computational complexities of the PSD model considered, good results were obtained using a gradient-based optimisation routine. The effectiveness of VAc and surfactant feed rates as decision variables was demonstrated. The further addition of the duration of the time intervals of injection did not result in substantial improvements in the final solution. Multi-objective ϵ-constrained strategies were capable to find pareto optimal recipes for PSD control. A restriction of the number of manipulated variables to only the feed rates of surfactant resulted in a loss of independent control of the nucleation and growth rates.

Future work will involve the experimental validation of the optimal recipes calculated, the implementation of batch-to-batch strategies to account for model uncertainties and process disturbances, and the application of the illustrated optimisation strategies to system using ionic surfactants and thermal initiators.

Acknowledgments

The authors acknowledge financial support from the EPSRC (Engineering and Physical Sciences Research Council)[Grant: GR/S94124/01].

References

1. P. F. Luckham and M. A. Ukeje, *J. Colloid Interface Sci.* **220**, 347 (1999).
2. F. J. Doyle III, M. Soroush and C. Cordeiro, *AIChE Symposium Series: Chemical Process Control VI, CACHE, New York: AIChE*, 290 (2002).
3. C. D. Immanuel, C. F. Cordeiro, S. S. Sundaram, E. S. Meadows, T. J. Crowley and F. J. Doyle III, *Comp. Chem. Eng.* **26**, 1133 (2002).
4. C. D. Immanuel, F. J. Doyle III, C. F. Cordeiro and S. S. Sundaram, *AIChE J.* **49**, 1392 (2003).
5. D. Ramkrishna, *Population Balances - Theory and Applications to Particulate Systems in Engineering*, Academic Press, San Diego, CA (2000).
6. C. D. Immanuel and F. J. Doyle III, *Chem. Eng. Sci.* **58**, 3681 (2003).
7. N. Sun, *Modelling and control of particle size distribution in emulsion polymerisation reactors*, M.Sc. thesis, Imperial College London, University of London, London, UK (2004).
8. F. J. Doyle III, C. A. Harrison and T. J. Crowley, *Comp. Chem. Eng.* **27**, 1153 (2003).
9. J. Flores-Cerrillo and J. F. MacGregor, *Ind. Eng. Chem. Res.* **42**, 3334 (2003).
10. C. D. Immanuel and F. J. Doyle III, *AIChE J.* **49**, 2383 (2003).
11. C. Sayer, G. Arzamendi, J. M. Asua, E. L. Lima and J. C. Pinto, *Comp. Chem. Eng.* **25**, 839 (2001).
12. O. Kammona, E. G. Chatzi and C. Kiparissides, *J. Macromol. Sci. - Rev. Macromol. Chem. Phys.* **C39**, 57 (1999).
13. C. D. Immanuel and F. J. Doyle III, *Chem. Eng. Sci.* **57**, 4415 (2002).
14. E. S. Meadows, T. J. Crowley, C. D. Immanuel and F. J. Doyle III, *Ind. Eng. Chem. Res.* **42**, 555 (2003).
15. D. Kozub and J. F. MacGregor, *Chem. Eng. Sci.* **47**, 929 (1992).
16. E. Saldivar and W. H. Ray, *AIChE J.* **43**, 2021 (1997).
17. NAG, *NAG (Numerical Algorithms Group) Fortran Library*, http://www.nag.com/numeric/fl/fldescription.asp (2005).
18. C. D. Immanuel and F. J. Doyle III, *Ind. Eng. Chem. Res.* **43**, 327 (2004).

APPLICATION OF PARALLEL ARRAYS FOR PARALLELISATION OF DATA PARALLEL ALGORITHMS

A. JAKUŠEV AND V. STARIKOVIČIUS

Department of Mathematical Modelling
Vilnius Gediminas Technical University, Saulėtekio ave 11
Vilnius, Lithuania, E-mail: alexj@fm.vtu.lt, vs@sc.vtu.lt

Data parallel algorithms are very common in both optimisation and process modelling problems. The size of problems solved by such algorithms may be significantly increased and absolute computation time may be reduced by using parallel computing. Parallel arrays is C++ library designed to simplify parallelisation of data parallel algorithms, using principle similar to High Performance Fortran. The algorithm must be implemented using special arrays instead of native C/C++ ones. Application of the library to parallelisation of porous media modelling algorithm and image smoothing is also described.
Keywords: parallel arrays, C++ parallelisation, data parallel algorithms.

1. Introduction

Data parallel algorithms are used very often to solve real-world tasks. An example of such task could be a modelling of porous media processes or image smoothing. Such algorithms are known to manipulate large amounts of data and perform large amount of computations. There are many ways to improve solver's implementation so that it can solve bigger problems in a reasonable amount of time. Faster and more memory-saving numerical methods should be applied, but using more powerful hardware is also an option.

While the power of modern PC is constantly growing, it is growing not enough to cover the demand for increasing size of problems to solve. In such cases, parallel computing may be the answer. Not that parallel computing only gives access to increasing computational resources, but it may also be economically effective, comparing to the cost of more powerful single PC.

The major difficulty in using parallel computers, however, is that writing a parallel program (or parallelising existing sequential code, that is abundant everywhere), requires the knowledge of special methods and tools, which is not trivial to be mastered.[1] Hence the main obstacle in the spreading of parallel computing is the lack of specialists who may create parallel software.

One of the ways to improve the situation is the creation of tools to simplify the above mentioned task of algorithm parallelising. The authors of this paper have analysed already available tools and created a new one—parallel array library ParSol. This library may help parallelising algorithms with parallel data implemented using `C++`. The examples of using the library for parallelising real-world modelling tasks are also described in this paper.

2. Parallel Programming Tools and Standards

There are several parallel programming standards and tools available, such as PVM (Parallel Virtual Machine[a]), MPI (Message Passing Interface), HPF (High Performance Fortran) or OpenMP application program interface[b]. We'll discuss MPI and HPF here, because these tools are used in this paper, and also an OpenMP, as a popular alternative for data-parallel algorithm parallelisation.

MPI (*Message Passing Interface*) is a standard for a `C/C++` or Fortran libraries.[2] It is wide spread, has lots of implementations on different platforms, both commercial and free. However, it is quite complicated, and parallelising programs using MPI is a tedious process.[3]

HPF (*High Performance Fortran*) is an extension of Fortran language standard. HPF is well suitable for algorithms with parallel data. If program is written in standard Fortran following some simple rules (considering the usage of Fortran arrays), then it may be parallelised just by adding several directives, describing such things as processor topology.[1] The drawbacks of HPF are diminishing popularity of Fortran language and the need to develop separate HPF compiler.

OpenMP is an Application Program Interface (API) that supports multi-platform shared-memory parallel programming in `C/C++` and Fortran on all architectures, including Unix platforms and Windows NT platforms. OpenMP is a portable, scalable model that gives shared-memory parallel programmers a simple and flexible interface for developing parallel applications for platforms ranging from the desktop to the supercomputer. However, OpenMP is not designed for the usage on distributed memory systems, such as PC clusters.

[a]Visit http://www.csm.ornl.gov/pvm/pvm_home.html for more information
[b]Visit http://www.openmp.org/ for more information

3. ParSol Features

The aim of ParSol is to bring HPF parallelisation simplicity to `C++` language, using popular parallelisation standards. Hence, the current ParSol library features are:

- Created for `C++` programming language;
- Based on HPF ideology;
- The library heavily uses such `C++` features as OOP (Object Oriented Programming) and templates;
- Only standard `C/C++` features are used, to improve portability. While using additional packages, such as OOMPI (Object Oriented MPI[c]), would help library developer, it would put additional requirements on the user system, making the task of using the library more difficult;
- Currently, MPI 1.1 standard is used to implement parallelisation.[2,3] This standard is implemented on large variety of platforms, while newer MPI version, 2.0, is not so widespread, or sometimes implemented partially;
- ParSol currently is open source library. The source code may be obtained from authors upon request.

Due to compliance to general standards, ParSol is expected to be used on wide variety of platforms. Currently ParSol has been tested on the following configurations:

- Microsoft Windows OS (98 and XP), Microsoft Visual `C++` 6.0 compiler, MPICH[d] 1.2.5 MPI library[e];
- Linux OS (kernel 2.4.21), gcc compiler version 3.2.3, LAM[f] MPI library version 7.1.2 (`http://vilkas.vtu.lt`);
- IBM SP4 supercomputer, VisualAge `C++` compiler version 5, IBM MPI library (`http://www.cineca.it`).

At present, ParSol may be used for parallelisation of data-parallel problems with explicit schemes.

---

[c]Visit http://www.osl.iu.edu/research/oompi/ for more information
[d]Visit http://www-unix.mcs.anl.gov/mpi/mpich/ for more details
[e]The library was tested with multi-process emulation on single PC provided by MPICH. While this should ensure that ParSol works correctly on this platform, no real experiments to estimate parallelisation efficiency could be conducted
[f]Local Area Multicomputer—visit http://www.lam-mpi.org/ for more details

ParSol is not the only solution for data-parallel algorithm parallelisation. Another popular standard, also designed for `C/C++`, is OpenMP. The user could choose ParSol over OpenMP if

- The user has strong background in HPF programming, and would like to reuse her experience in `C/C++`invironment;
- The user wants to be able to execute her parallel code on every platform where MPI is supported.

The fact that ParSol relies upon MPI for communication means that its effectiveness depends on the effectiveness of underlying MPI implemeneta-tion. For example, if MPI implementation on shared-memory system uses shared memory for message passing (as many good MPI implementations do), then the speed of ParSol library on shared memory systems may be similar to OpenMP.

4. ParSol Structure and Usage

The main elements of the library are:

Sequential array classes. These are the classes to be used instead of native `C/C++` arrays even in sequential programmes. No MPI or other libraries, except ParSol itself, is necessary to use sequential classes. Comparing to native `C/C++` arrays, ParSol sequential arrays have a number of advantages for programming mathematical algorithms, such as virtual indexes, built-in array operations, automated management of dynamically allocated memory.

Parallelisation and parallel array classes. If parallel arrays are to be used in place of sequential ones, it is natural to make them the descendants of appropriate sequential arrays, adding parallelisation code to the sequential array functionality. However, parallelisation is similar for different kinds of arrays. So parallelisation code is localised in class `PS_ParArray`, and is used in parallel array classes by multiple inheritance.

The parallelisation method is graphically displayed on Fig. 1. It consists of the following steps (given that processes are in proper order):

(1) Determine the part of sequential array that belongs to given process;
(2) Determine the neighbour processes that will participate in information exchange;
(3) Determine the amount of data to be exchanged with every neighbour process;
(4) Exchange information with neighbours, when required.

Topology classes. The purpose of these classes is to ensure that all processes are in proper order for parallel array functionality. In HPF, this functionality is performed by special directives.

All the general code resides in `PS_CustomTopology` class. As with sequential array classes, there are also descendants for some special cases (`PS_{1,2,3}DTopology`), which provide end user with more friendly interface.

Stencil classes. Stencil is determined depending on what computational scheme is used. Based on stencil, different amount of information needs to be exchanged among neighbours (see Fig. 1 (b)). Hence, stencil information is required for parallel arrays to operate properly.

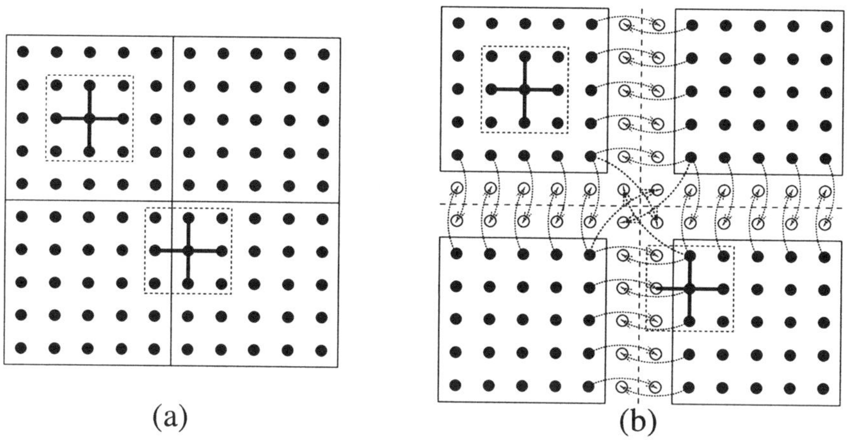

Figure 1. Transition from sequential (a) to parallel (b) array.

To use ParSol, a programmer must develop his/her sequential application in the same way as without ParSol, only using ParSol arrays wherever computational data is stored. Another requirements are to specify the stencil, make algorithm independent on the order in which array points are processed and use array operations provided by ParSol wherever possible. The last one may also be called an advantage, because it frees programmer from implementing simple tasks, allowing to concentrate on problem solving, and makes code cleaner.

The parallelisation of such a program takes the following steps:

(1) replace includes of sequential headers with parallel ones (ex.:

`PS_CommonArray.h` to `PS_ParallelArray.h`);
(2) replace sequential classes with their parallel analogy in variable declarations only;
(3) Add MPI initialisation code (one line in the beginning of the program);
(4) Add topology initialisation code (in its simplest, one line in the beginning of the program);
(5) Specify when array neighbours should exchange data.

Finally, MPI library should be linked during building process.

5. Application of ParSol to Porous Media Problem Solver

A porous media consists of solid phase and void spaces. The spaces may be filled with various gaseous and fluid phases. In order for the media to be considered as porous, the following requirements must be fulfilled:

(1) The dimensions of the void space must be small enough so that the fluid flow is controlled by adhesive and cohesive forces;
(2) The dimensions of the void space must be large compared to the mean free path length of the fluid molecules;
(3) The void spaces of the porous media is interconnected.

In such a case porous media may be modelled by the following equations:

$$\frac{\partial\left(\Phi\rho_\alpha S_\alpha\right)}{\partial t} + \nabla\left\{\rho_\alpha \mathbf{u}_\alpha\right\} = \rho_\alpha q_\alpha \ , \tag{1}$$

$$\mathbf{u}_\alpha = -\frac{k_{r\alpha}}{\mu_\alpha}\mathbf{K}\left(\nabla p_\alpha - \rho_\alpha \mathbf{g}\right) \ , \tag{2}$$

$$p_{c\beta\alpha}\left(\mathbf{x},t\right) = p_\beta\left(\mathbf{x},t\right) - p_\alpha\left(\mathbf{x},t\right), \ \beta \neq \alpha \ , \tag{3}$$

$$\sum_\alpha S_\alpha = 1 \ , \tag{4}$$

where (1) is mass conservation law for every phase α, (2) is Darcy law for every phase α, (3) is capillary pressure for all phase pairs (α, β), and (4) is for saturations.

To solve this equation system, *global pressure formulation* approach is used. When using this approach, we have less connected equations and input values are smoother.[4]

Results of using ParSol on implicit problems are presented below. Here, p is the number of processors, N is the task size, T_p is an execution time, S_p is a speedup and E_p is a parallelisation efficiency. Parallelisation efficiency on SP4 supercomputer is very high, as it may be seen in Table 1. The super-speedup effect is due to hardware pecularities of a given system—it processes smaller arrays more efficiently. The following CPU times $T_1(N)$ (in s) were obtained for the sequential algorithm

$$T_1(160) = 64.97, \quad T_1(320) = 241.4, \quad T_1(480) = 281.9.$$

Table 1. Implicit nonlinear diffusion algorithm on SP4.

p	$S_p(160)$	$E_p(160)$	$S_p(320)$	$E_p(320)$	$S_p(480)$	$E_p(480)$
2	2.42	1.211	2.78	1.392	2.38	1.189
4	5.04	1.260	5.98	1.495	4.41	1.102
6	7.03	1.172	8.97	1.495	6.58	1.097
8	8.56	1.070	11.30	1.412	8.69	1.086
16	13.45	0.841	23.44	1.465	17.15	1.072

Parallelisation efficiency on PC cluster (Table 2) is worse because of higher communication costs, but communication costs impact is comparatively low because implicit problems are computationally intensive. The following CPU times $T_1(Iterations \times Size)$ (in s) were obtained for the sequential algorithm on PC cluster

$$T_1(188 \times 100) = 24.10, \quad T_1(350 \times 200) = 366.54.$$

Table 2. 3D Poisson equation, using CG method and 7-point stencil on PC cluster.

p	$S_p(188 \times 100)$	$E_p(188 \times 100)$	$S_p(350 \times 200)$	$E_p(350 \times 200)$
2	1.82	0.911	1.98	0.988
4	3.63	0.906	3.86	0.965
8	5.97	0.747	7.11	0.888

6. Application of ParSol to Image Smoothing

Image smoothing has wide range of applications. Its mathematical models are based on partial differential equations approach. It is well known that very popular image filters are obtained by convolution with Gaussian function G_σ of increasing variance. The application of such a filter is equivalent to solving a linear parabolic problem

$$\begin{cases} \dfrac{\partial u(X,t)}{\partial t} = \sum\limits_{i=1}^{2} \dfrac{\partial^2 u(X,t)}{\partial x_i^2}, \quad X = (x_1, x_2) \in Q, \\ u(X,0) = u_0(X), \end{cases} \tag{5}$$

where u_0 is an initial image and $t^{1/2}$ defines the the variance of the Gaussian function. To have selective diffusion, a following modification may be used:[5]

$$\begin{cases} \dfrac{\partial u(X,t)}{\partial t} = \sum\limits_{i=1}^{2} \dfrac{\partial}{\partial x_i}\Big(g\big(|\nabla G_\sigma * u(X,t)|\big), \dfrac{\partial u(X,t)}{\partial x_i}\Big) + f(u_0 - u), \\ \partial_\nu u = 0, \quad (X,t) \in \partial Q \times (0,T], \\ u(X,0) = u_0(X), \quad X \in Q. \end{cases} \tag{6}$$

Usually a discrete image is given on a structure of pixels with rectangular shape. This fact defines a discrete space mesh, and we also introduce a uniform time grid. Then the following discretization scheme may be obtained using finite volume method:

$$\partial_t U_{ij}^{n+1} = \sum_{\alpha=1}^{2} \partial_{x_\alpha}^{+} \left(a_\alpha(U_{ij}^n)\, \partial_{x_\alpha}^{-} U_{ij}^n\right) + f\left(u_{0,ij} - U_{ij}^n\right), \tag{7}$$

This is explicit scheme, and it is stable only when $\tau \leqslant ch^2$. Always stable implicit scheme may also be obtained.

The results of using ParSol for parallelisation of explicit algorithms are presented in Table 3 for PC cluster. The filtration problem was solved till time moment $T(N)$ and the following CPU times $T_1(N)$ (in s) were obtained for the sequential algorithm

$$T(160) = 0.1, \quad T_1(160) = 213.3, \quad T(240) = 0.03, \quad T_1(240) = 332.8,$$
$$T(320) = 0.01, \quad T_1(320) = 361.6.$$

For SP4 supercomputer, results are presented in Table 4. The following CPU times $T_1(N)$ (in s) were obtained for the sequential algorithm

$$T_1(80) = 57.24, \quad T_1(160) = 471.2, \quad T_1(320) = 770.4.$$

We may notice bigger negative communication costs impact on PC cluster, due to the fact that explicit algorithms are computationally less intensive.

Table 3. The speedup and efficiency for explicit diffusion algorithm on PC cluster.

p	$S_p(160)$	$E_p(160)$	$S_p(240)$	$E_p(240)$	$S_p(320)$	$E_p(320)$
2	1.56	0.780	1.76	0.880	1.87	0.934
4	2.36	0.590	3.00	0.750	3.45	0.862
6	2.78	0.463	3.93	0.655	4.77	0.795
8	2.95	0.369	4.69	0.585	5.88	0.735
9	3.16	0.351	5.04	0.560	6.28	0.698
11	3.33	0.303	5.50	0.500	7.09	0.644
12	3.35	0.279	5.64	0.470	7.47	0.623
15	3.39	0.226	6.38	0.425	8.56	0.571

Table 4. The speedup and efficiency for explicit diffusion algorithm on SP4.

p	$S_p(80)$	$E_p(80)$	$S_p(160)$	$E_p(160)$	$S_p(320)$	$E_p(320)$
2	1.975	0.988	1.984	0.992	2.004	1.002
3	2.794	0.931	2.950	0.985	2.970	0.990
4	3.741	0.935	3.928	0.982	3.986	0.996
6	5.168	0.861	5.463	0.910	5.916	0.986
8	6.766	0.846	7.293	0.911	7.831	0.979
9	6.784	0.754	7.604	0.845	8.467	0.941
12	8.701	0.725	10.19	0.849	11.216	0.934
16	10.84	0.677	12.75	0.797	15.041	0.940
24	14.18	0.591	18.24	0.760	21.961	0.915

7. Conclusions

Data parallel algorithms are often used in modelling of various processes. One of the ways to quickly and easily increase the size of the problem and/or reduce the computational time for data-parallel algorithms implemented in C++ is to parallelise them using ParSol parallel array library developed by the authors.

The idea behind ParSol is similar to HPF, and ParSol may be used on a wide variety of platforms, where MPI 1.1 standard is implemented and C++ compiler is available. These and other features make ParSol a tool to consider for parallelisation of data-parallel algorithms, in the same class with such well known standard as OpenMP.

ParSol has already been tested on such problems as modelling of porous media problems (implicit algorithms) and image smoothing (explicit algorithms). In both cases ParSol provided very good parallelisation efficiency, considering algorithms pecularities and used hardware.

References

1. R. Čiegis, *Parallel Algorithms*, Technika, Vilnius (2001) (in Lithuanian).
2. Message Passing Interface Forum, *MPI: A Message-Passing Interface Standard*, June 12, (1995).
3. W. Gropp, E. Lusk and A. Skjellum, *Using MPI. Portable Parallel Programming with the Message-Passing Interface*, The MIT Press (1999).
4. A. Jakušev and V. Starikovičius, Multiphase Flow problem solver and its application for multidimentional problems, *Lithuanian Mathematical Journal* **44**, 634–638 (2004).
5. P. Perona and J. Malik, Scale space and edge detection using anisotropic diffusion, In: *Proc. IEEE Computer Society Workshop on Computer Vision* (1987).

CAD GRAMMARS: EXTENDING SHAPE AND GRAPH GRAMMARS FOR SPATIAL DESIGN MODELLING

P. DEAK, C. REED AND G. ROWE

School of Computing, University of Dundee, UK

Shape grammars are types of non-linear formal grammars that have been used in a range of design domains such as architecture, industrial product design and PCB design. Graph grammars contain production rules with similar generational properties, but operating on graphs. This paper introduces CAD grammars, which combine qualities from shape and graph grammars, and presents new extensions to the theories that enhance their application in design, modelling and manufacturing. Details about the integration of CAD grammars into automated spatial design systems and standard CAD software are described. The benefits of this approach over traditional shape grammar systems are also demonstrated.

1. Introduction

The aim of the Spadesys project is to investigate how spatial design and modelling can be automated in a generalised way, by connecting similar concepts across the various design domains and decoupling them from the intelligent design process. The core functionality of the system is based on a generative approach to design generation using CAD grammars. The initial part of this paper provides a brief description of shape and graph grammars – two of the base concepts behind CAD grammars. Afterwards, the extensions proposed by CAD grammars are introduced, and the benefits of their use in process plant layout design explained.

2. Shape Grammars

Shape grammars have proved to be applicable in a range of different design domains from camera to building design,[1] which sets them as an appropriate technique to further the goals of generalised design. They employ a generative approach to creating a design using match and replace operations described by a grammar rule set for a domain. There are, however, a number of limitations of shape grammars:

- Engineering domains will have a large set of inherent domain requirements, and each specific design to be generated will have a large set of problem specific requirements and constraints related to that instance. Creating a grammar rule set that contains the maximal amount of domain knowledge, while remaining flexible and adaptable enough to fulfil the greatest number of designs can result in a large or complex grammar rule set.

- Communicating grammar effectively is difficult; justification for individual grammar rules can be difficult to provide, as they may not have a direct significance on a design, instead playing a linking role where they prepare parts of the design for further grammar rules to work on. This can make maintenance, and understanding of the grammar by anyone who was not involved with its creation difficult.
- In order to use shape grammars in an automatic design generation scenario in most engineering domains, the grammar has to be very detailed and complete, and prohibit the introduction of flaws into the design.
- It is difficult to verify a grammar. A recursive rule set can define an infinite space of possible solutions, and can therefore contain designs that may be flawed in ways that were not anticipated by the grammar designer.
- Current shape grammar implementations do not make it possible to express connectivity; if two line segments in a design share a common endpoint, it is not possible to show whether they are segments of a logically continuous line, or two unrelated lines which happen to be coincident.
- It is Difficult to create a 'designerly' grammar, where the order and application of rules proceeds and a way that makes sense to the user.

3. Graph Grammars

Graph grammars[2] consist of production rules to create valid configurations of graphs for a specific domain. They have been successfully employed in designing functional languages[3] and generating picturesque designs.[4] Graph grammar rules contain the match and replace operations for nodes and edges in a network.

There is generally no spatial layout information associated with the nodes and edges; the only relevant data is the types of nodes and edges, and the information about the connections between them. It is therefore difficult to model spatial and graphical designs with graph grammars alone. A desirable feature with graph grammars is that the application of grammar rules keep the design connected as the network is increased.

4. Shapes and Graphs

In typical CAD applications, some of the primitives used to model designs are vertices (points in 3D space), edges (lines connecting points), and faces (enclosed polygons made by edges). This has proven to be an effective way of representing many types of spatial data, as it allows for a range of editing and analytical operations to be applied to a model. Vertices represent a sense of connectivity between lines. This makes it helpful to display and edit designs and express relationships between lines. Traditional shape grammar systems are not able to deal with CAD primitives directly, as the only components that can be present are shapes or volumes. A CAD design would first have to be represented as shapes or volumes only. Clearly, it would be desirable if the representation does not have to be altered from the one used in CAD software.

There is a clear correlation between these CAD elements and graphs. A design represented using CAD elements can be seen as a graph, with the vertices being the nodes of the graph and lines being the arcs or edges. A CAD design is more complex however, and contains more information, as not only the presence of nodes and arcs, but also their positions and lengths are relevant. Graph grammars have been used in a similar way to shape grammars to design graphs, and an advantage of graph grammars is that there is a sense of connectivity between the elements.

In the Spadesys system, one of the core ideas is to combine shape grammars with graph grammars, inheriting the beneficial features of both concepts. Additionally, in Spadesys there are a number of extensions and new possibilities which are not found in any other shape or graph grammar system. "CAD grammars" are thus an amalgam of the two systems, and inherit benefits from both. In order to address remaining limitations, a number of extensions are proposed, and their implementation in Spadesys is discussed.

5. CAD Grammar Fundamentals

Rules in CAD grammars are comprised of two parts, the match shape, which is a specification of the shape to be matched, and the replace shape, which is the shape to replace the specified match shape. The design shape is the specification of the current design that is being generated. The matching algorithm looks to find occurrences of the match shape within the design shape, and replace those configurations with the replace shape.

The basic elements for shapes in a CAD grammar system are *points* and *lines*. *Points* are objects which have the numerical parameters x, y (and z in a 3D implementation). Lines are represented by references to two points; $p0$ and $p1$. It is important to consider points and lines as objects; as there may be multiple points with the same parameters, but are distinct entities.

Connectivity among two lines can be represented by the two lines sharing a common point instance. In CAD grammars it is important to be able to make this distinction in the design shape and the match/replace shape. The usefulness of this feature can be seen in instances where two lines happen to appear to share an endpoint, but they are not intended to be logically continuous with regards to the grammar matching algorithm.

Figure 1 shows an example of the connectivity features of CAD grammars. Continuous, connected lines are shown with LineA(Point1, Point2) and LineB(Point2, Point3). Non-connected lines: LineC: Point4, Point5 and LineD: Point6, Point7.

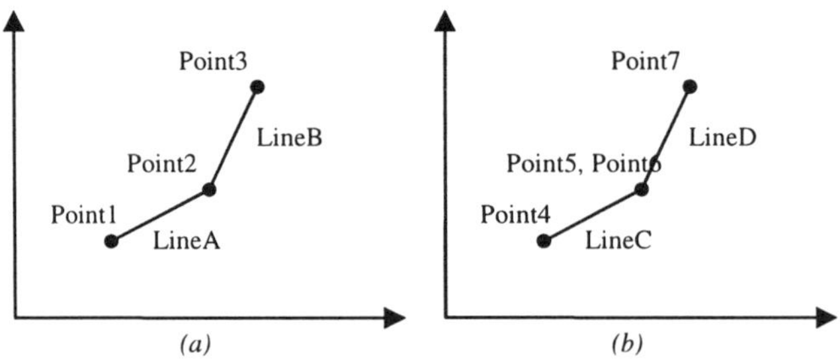

Figure 1. Line Connectedness.

In Fig. 1(a), the two line segments are connected, which can be seen by the use of only three point instances, with Point2 being shared by both line segments. Figure 1(b) shows spatially identical, non-connected lines, with each line having unrelated point instances.

Similarly, intersecting lines do not logically subdivide into four line segments, as is often the case in traditional shape grammar systems. That intention can be defined implicitly by the presence of a point at the intersection, with the four line segments connected to it. The reason for this is that there are many cases when the result of applying certain grammars results in lines intersecting, but it is not the intention of the grammars to have the intersection produce corners which are matched by other grammar rules. This can prevent accidental, unintended matches in further operations on a shape. For example, the match shape in Fig. 2(a) would successfully match the design shape in Fig. 2(b), but not that in Fig. 2(c).

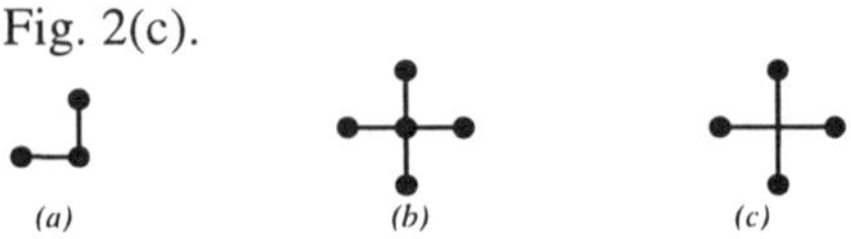

Figure 2. Matching connected lines.

6. Extension 1: Length and Angle Constraints

Parametric shape grammars[5] allow specification of variable parameters in shape grammars. In Spadesys's CAD grammar, this idea is taken a step further to allow much more customization in the expression of match shapes. Every line in a match shape can have a length constraint. This length constraint is evaluated with the line that is to be matched in the current design when running the matching algorithm.

With regards to many engineering design domains, there may be a need to specify exact line sizes in the match shape, which will result in lines only of that exact length being matched. In CAD grammars, if the length constraint for a line is an exact value, then that line will only match lines of that value. This allows match shapes to be drawn inexactly when actual values are known for the line lengths. Similarly, the length constraint may be specified as a range such as 4-10, in which case all lines of length between 4 and 10 will be matched. Logical operators can be used within the length constraint to allow further control on matching; for example we want to match lines of length 7 or 15, we can set its length in the match shape to 7 | 15. Similar constraints can also be applied to angles between lines, to provide similar flexibility with regards to appropriate angles too.

When the length constraint is set to *proportional*, the behaviour is similar to traditional shape grammars, where any line length will match, provided that all the lines which were matched have the same proportions as the lines in the match shape, making the scale of the match shape irrelevant. When the length constraint is set to *length*, then the exact length of the line is used, as it is shown graphically in the match shape. This is different from exactly specified lengths, as they may be a completely different size from the physical length of the line in the shape.

Due to the complete scripting system embedded within Spadesys, complex mathematical operations can be also used in the length constraint.

7. Extension 2: Modification

Shape grammars (as well as all formal grammars) operate using match and replace operations only. When the aim of a grammar rule is to modify a feature, it is achieved by having a similar match and replace shape, which vary in terms of the intended modification. In standard shape grammars, this approach is fine, since there is no difference between actually modifying the matched shape's elements in the current design, or simply removing it and inserting a new shape as desired. However in CAD grammars there can be meta-information associated with the lines and points in a design, which in many cases would need to be retained.

The most important part of the meta-information of a line is its connectedness; i.e. which other lines it is connected to. It is necessary to be able to state in a grammar rule whether the elements in the current shape should be replaced by new instances of the elements in the replace shape, or whether they should be modified as stated by the elements in the replace shape. The effect of this idea in practice is that grammar rules can not only match and replace, but they can also *modify*. This means that there can be two grammar rules that look identical with regards to the lines and points, but create a completely different result when applied. This is unlike the effect of modification that can be achieved using only match and replace, as seen in the following examples.

The grammar rule in Fig. 3 is designed to stretch the match shape regardless of context.

Figure 3. 'Stretch' shape grammar rule.

When applied traditionally to the following example, unintended results are produced, so that the design shape in Fig. 4(a) changes to the shape in Fig. 4(b), rather than what was intended: Fig. 4(c).

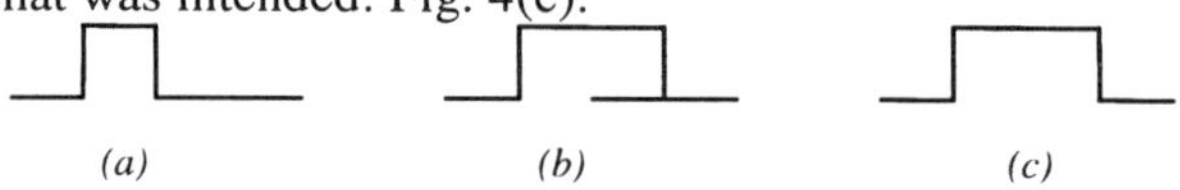

Figure 4. Tradition application of rule.

To get the intended result with a traditional shape grammar approach, there would need to be a larger, more complex grammar that takes into account all possible contexts of the original match shape, and modifies the effected portions of the design shape separately. In Spadesys, the above grammar rule from Fig. 3 would be represented as the rule in Fig. 5.

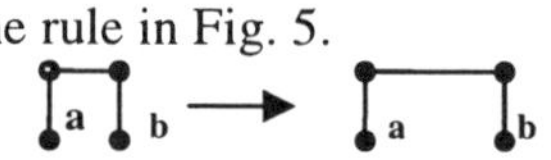

Figure 5. Connectedness in matching.

This modification ability is currently implemented using a tagging system. The points in the match and replace shape can be tagged with labels (strings) to signify their correspondence. In Fig. 5, the 'a' and 'b' labels associated with the points represent their tags. If a point in the replace shape has the same tag as a point in the match shape, then the matched point in the design shape will be modified to spatially match the replace point, as opposed to removing it and replacing it with a new point. This ensures that the connectivity of the point in the design shape is maintained after the replace operation, and gives the effect of modifying the shape, as opposed to deleting and inserting portions.

8. Extension 3: Line types

In current shape grammar systems, non-terminals are generally represented by a combination of terminals that is unlikely to be found elsewhere in the shape (only in the parts where it is intended to be a non-terminal). This requires complicating the design, and is not safe or efficient. Colour grammars[6] can be used to improve this method, but Spadesys takes the idea a step further by introducing line types. The lines in a match shape have a type rather than a colour. Types are hierarchically structured entities in the same sense as classes and subclasses are in programming languages. The base type is *Line*, and all other line types derive from it. Due to the polymorphic nature of types, if a line in a match shape is of type *Line*, then it will match any type of line in the current design (provided the length constraint is also met).

Generative design often takes place in phases,[1] by gradually lowering the level of the solution from a high-level/abstract design to a low-level/complete design, until it satisfactorily represents the requirements. For example in architecture, the solution can initially start off as a grid of squares covering the approximate layout of the intended design. Applying an initial grammar set in the first phase will add some temporary walls to outline a basic conceptual layout. The next phase can add more detail on the shape of the walls, and position them adequately. Further phases may add additional details such as doors or windows, and so on. By annotating the lines in grammars with their types, we can show clearly which grammars should be applied at the first phase (by setting the match shapes lines to type *grid*) and what phase it will prepare its results for (by setting the replacement shapes lines to type *basicwall*). This opens up more flexible approaches with regards to the progression of the shape generation. One half of the building can be generated right up to the windows and doors phase, and once satisfactory, the other half may be worked on without interference. This region based workflow may be more appropriate in some cases than a phase based one.

Grammar interference is also removed, and the grammars from different phases do not have to be handled separately. A grammar rule will only be applied where it is intended to be applied, on the types of lines it is intended to be applied.

A grammar rule becomes self documenting to an extent, as the line types describe when and where it is applied, and more accurately shows what the designer is trying to achieve with the grammar rule.

9. Partial Grammars

Spadesys attempts to drive the use of partial grammars as a way to tweak and modify designs in a clear and simple way. Formal grammars theory states that all valid designs must derive from the grammar rule set in play. However, when the aim is to modify existing designs with new features, it may be inefficient to determine their grammar rule set and modify it in a suitable way so that when the design is re-generated it contains the intended changes. It may be simpler having a partial grammar containing only the rules for the new features, and applying that to modify the design. A partial grammar is a reduced set of grammar rules with the intent to modify existing designs, rather than generate a complete design from nothing. For example, given the existing design in Fig. 6(a), the aim is to round off the edges to produce Fig. 6(b).

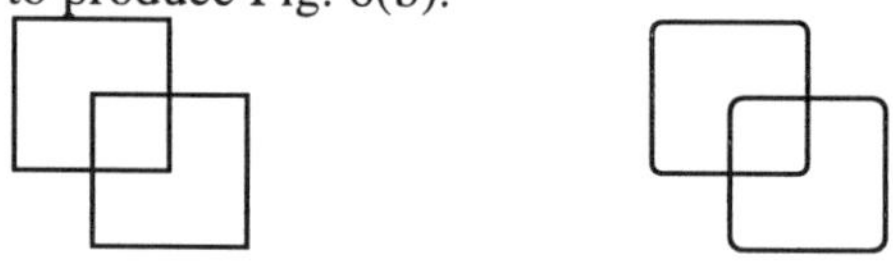

(a) *(b)*

Figure 6. An example of modification.

The complete grammar would either have to contain all the rules to produce the source shape with the addition of rules to perform the modification, or the rules would have to be modified so that the intended design is produced directly. Either way requires the original grammar, which may not exist and can be difficult to derive. In Spadesys, the grammar rule similar to the one in Fig. 7 can be directly applied to any design shape.

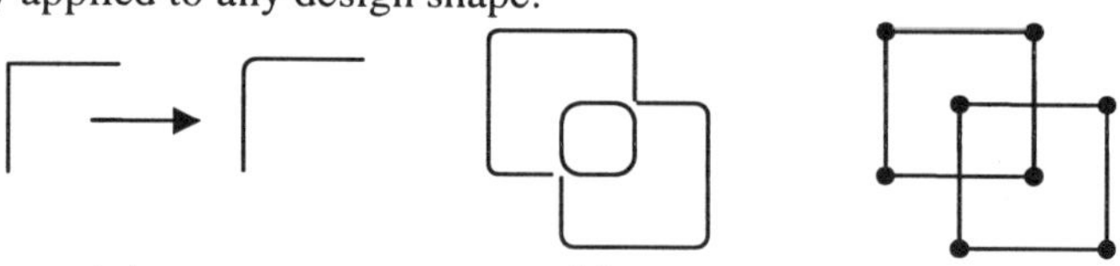

(a) *(b)* *(c)*

Figure 7. Rounding.

The application of the rule in Fig. 7(a) on the design shape in Fig. 6(a) demonstrates another useful feature of CAD grammars that derives from the extended connectivity features. Without the connectivity information in the design shape, automatic application would be difficult, as unexpected results can be produced, such as the one in Fig. 7(b). With the CAD grammar elements, the initial design shape would be represented as shown in Fig. 7(c), with the intended corners specified. This way there can be control over how intersections are treated, and this control is given to the user, rather than determined by the implementation.

Partial grammars may also be used as the basis for generating the design. In an architectural example, a grammar for a house may be designed in a way, that it is incomplete, and cannot generate a design on its own. However, when provided with a building outline as the current design, it can generate the remainder of the house.

The modification features of CAD grammars as described above in extension 2 are very assistive to the idea of using partial grammars. The modifications can be represented in a compact, context free way by being able to preserve connections between lines and therefore modify the surrounding context suitably. The length constraints feature as described in extension 1 is also a valuable feature for such situations, because a single grammar rule becomes more flexible and can apply to more varying configurations.

10. Application Domain – Process Plant Layout Design

One of the intended domains for evaluation of the Spadesys system is process plant layout design.[7] Process plant design is a complex field with several conflicting factors, and a large number of constraints controlling the design generation. The layout design is the phase that is to be performed by Spadesys, with results of the earlier phases being used as inputs to the design generation. These earlier phases will have determined the equipment and their relationships among them, as well as the overall layout and size constrains imposed by the plant. These requirements have to be represented in the problem code, so the design layout generation can use analyze the constraints and commence layout generation.

Using a CAD grammar based automated design generation system for a domain such as plant design has several benefits:

- One of the main factors in design generation is hazard prevention.[8] The grammar rules to generate the layout can be designed in a way to prevent many sources of hazard relating to spatial configurations. Since the grammar rules are the building blocks the generate the design, these *blocks* can be made so that they only fit together safely.
- Cost and space reduction is supported by the iterative approach provided by a system such as Spadesys. After selecting the most appropriate candidates from an initial set of designs proposed by the system, each can be further branched out or refined further to increase the suitability of the results. This allows a controlled level of user interaction to ensure the users are content with the way the designing is being performed.

11. Conclusion

CAD grammars provide a more flexible approach to the applications of shape grammars. The enhanced matching features allow the construction of smaller grammar rule sets. Grammar rules can adapt and match a larger number of relevant configurations due to the length and angle constraint features. The modification features, which allow grammars to directly modify designs introduces a new dimension to design generation.

The emergent features of shape grammars, where large complex designs can be generated from a few simple rules are still present; however the ability to define clear and predictable grammars is also enhanced. Using the *line-types* extension has not only functional benefits; in addition grammar rules can become self-documenting, with their features and intentions made clear by the visible type name of every line in the match and replace shapes.

The polymorphic nature of line-types provides more power to the grammar designer, by being able to further specify the intentions of grammars. This allows for the creation of more *'designerly'* grammars, where the design process can flow consistently with the design intention of the grammar designer. The use of CAD grammars allows a homogeneous model to be used for the design representation and the tools used to generate it.

References

1. G. Stiny and W. J. Mitchell, The palladian grammar, *Environment and Planning B* **5**, 5-18 (1978).
2. D. Plump, *Handbook of Graph Grammars and Computing by Graph Transformation* Vol. 2, Chap. 1, 3–61, World Scientific, Singapore (1999).
3. E. Barendsen and S. Smetsers, *Graph Rewriting Aspects of functional Programming*, 17, Chap. 2, pp. 62–102 (1999).
4. F. Drewes, Tree-based picture generation, *Theoretical Computer Science* **246**(1), 1-51 (2000).
5. G. Stiny, Introduction to shape and shape grammars, *Environment and Planning B* **7** (1980).
6. T. W. Knight, Color grammars: designing with lines and colors, *Environment and Planning B: Planning and Design* **16**, 417-449 (1989).
7. D. I. Patsiatzis, G. Knight and L. G. Papageorgiou, An MILP Approach to Safe Process Plant Layout. *Chemical Engineering Research and Design, Part A* **82**, 579-586 (2004).
8. M. Goring and H. G. Schecker, HAZEXPERT: An integrated expert system to support hazard analysis in process plant design, *Computers Chemical Engineering* **17**, 429-434 (1993).

MULTIDIMENSIONAL SCALING USING PARALLEL GENETIC ALGORITHM

A. VARONECKAS

Vytautas Magnus University, Vileikos 8
Kaunas LT-44404, Lithuania, E-mail: a.varoneckas@if.vdu.lt

A. ŽILINSKAS AND J. ŽILINSKAS

Institute of Mathematics and Informatics, Akademijos 4
Vilnius LT-08663, Lithuania, E-mail: antanasz@ktl.mii.lt, julius.zilinskas@mii.lt

Multidimensional scaling is a technique for visualization of multidimensional data. A difficult global optimization problem should be solved to minimize the error of visualization. Parallel genetic global optimization algorithm for multidimensional scaling is implemented to enable solution of large scale problems in acceptable time. Results of visualization using high performance computer and cluster of personal computers are presented.

1. Introduction

Multidimensional scaling (MDS) is an exploratory technique for data analysis.[1,2,3] The points $\mathbf{x}_i = (x_{i1}, \ldots, x_{im}), i = 1, \ldots, n$ representing n objects in m-dimensional embedding space should be found fitting pairwise distances of points to given pairwise dissimilarities of the objects (δ_{ij}). Under mild conditions there exist points in $(n-1)$-dimensional space whose pairwise distances are equal to dissimilarities.[2] However such points do not exist in general case when dimensionality of embedding space m is small. The two-dimensional embedding space ($m = 2$) is of special interest, when MDS is used to visualize multidimensional data.

The implementation of a MDS method is reduced to minimization of a fitness criterion, e.g. the so called *STRESS* function:

$$S(X) = \sum_{i<j}^{n} (d(\mathbf{x}_i, \mathbf{x}_j) - \delta_{ij})^2, \tag{1}$$

where $X = (\mathbf{x}_1, \ldots, \mathbf{x}_n)$; $d(\mathbf{x}_i, \mathbf{x}_j)$ denotes the distance between the points $\mathbf{x}_i$ and $\mathbf{x}_j$.

The distances may be estimated using different norms in R^m. Most often a Minkowski distance is used:

$$d(\mathbf{x}_i, \mathbf{x}_j) = \left(\sum_{k=1}^{m} |x_{ik} - x_{jk}|^r \right)^{1/r}. \tag{2}$$

The formula (2) defines the Euclidean distances when $r = 2$, and the city block distances when $r = 1$. The points $\mathbf{x}_i$ defined by means of minimization of (1) but using different distances in the embedding space can be interpreted as different nonlinear projections of the objects from the original to the embedding space. When objects of problem are defined by multidimensional points, dissimilarities can be found estimating pairwise distances of points in the original multidimensional space.

MDS is a difficult global optimization problem. Although *STRESS* is defined by an analytical formula, which seems rather simple, its minimization is difficult. The function normally has many local minima. The minimization problem is high dimensional: number of variables is $N = n \times m$. Non-differentiability of *STRESS* normally cannot be ignored, as *STRESS* is not differentiable when the points $\mathbf{x}_i$ and $\mathbf{x}_j$ are equal. However, at least *STRESS* is differentiable at local minimizer in case of Minkowski distances.[4]

2. Hybrid Global Optimization for Multidimensional Scaling

The hybrid algorithm combining evolutionary global search with efficient local descent is the most reliable though the most time consuming method for MDS with Euclidean distances.[5,6] Therefor a hybrid algorithm similar to one proposed by Mathar and Žilinskas[7] has been constructed. The pseudocode of the algorithm is outlined below.

The idea is to maintain a population of best (with respect to *STRESS* value) solutions whose crossover can generate better solutions. The size of population p is a parameter of the algorithm. An initial population is generated performing local searches from p starting points that are best (with respect to *STRESS* value) from a sample of N_{init} randomly generated points. The population evolves generating offsprings. Minimization terminates after predetermined computing time t_c.

The structure of the hybrid algorithm with parameters (p, N_{init}, t_c)

Generate the initial population:
    Generate N_{init} uniformly distributed random points.
    Perform search for local minima starting from the best p generated points.
    Form the initial population from the found local minimizers.
while not time-limit t_c exceeded
    Select two uniformly distributed random parents from a current population.
    Produce an offspring by means of crossover and local minimization.
    If it is better than the worst individual of the current population,
        then the offspring replaces the latter.

The upper level genetic algorithm ensures globality of search. Local descent at lower level ensures efficient search for local minima. From the point of view of evolutionary optimization the algorithm consists of the following "genetic operators": random (with uniform distribution) selection of parents, two point crossover, adaptation to environment (modeled by local minimization), and elitist survival. Interpreting the vector of variables in (1) as chromosome the crossover operator is defined by the following formula

$$X = arg\_min\_from((\hat{x}_{11}, ..., \hat{x}_{\xi_1 1}, \check{x}_{\xi_1+1 1}, ..., \check{x}_{\xi_2-1 1}, \hat{x}_{\xi_2 1}, ..., \hat{x}_{n1}), \\ (\hat{x}_{12}, ..., \hat{x}_{\xi_1 2}, \check{x}_{\xi_1+1 2}, ..., \check{x}_{\xi_2-1 2}, \hat{x}_{\xi_2 2}, ..., \hat{x}_{n2})), \quad (3)$$

where X is the chromosome of the offspring; $\hat{X}$ and $\check{X}$ are chromosomes of the selected parents; ξ_1, ξ_2 are two integer random numbers with uniform distribution over $1, ..., n$; and it is supposed that the parent $\hat{X}$ is better fitted than the parent $\check{X}$ with respect to the value of *STRESS*. $arg\_min\_from(Z)$ denotes an operator of calculation of the local minimizer of (1) from the starting point Z.

Direction set algorithm by Powell implemented by Press *et al.*[8] has been used for local search for MDS problems with Euclidean distances. Improved local search has been used for MDS problems with city block metric exploiting that in this case *STRESS* is a piecewise (over simply defined polyhedrons) quadratic function of X.[9] It has been shown that these local search strategies perform best in MDS with Euclidean and city block metrics.[10]

Parallel version of genetic algorithm with multiple populations[11] has been developed. Communications between processors have been kept to minimum to enable implementation of the algorithm on clusters of personal computers. Each processor runs the same genetic algorithm with different sequences of random numbers. This is ensured by initializing different seeds for random number generators in each processor. The results of different processors are collected when search is finished after predefined time.

To make parallel implementation as much portable as possible the general message-passing paradigm of parallel programming has been chosen. A standardized message-passing communication protocol MPI[12] is used for communication between parallel processors.

3. Experimental Investigation

To exclude the impact of number of objects and of used metric a relative error

$$f(X) = \sqrt{\frac{S(X)}{\sum_{i<j} \delta_{ij}^2}}, \tag{4}$$

is used for comparison. Performance of the hybrid global optimization algorithm for multidimensional scaling in visualization of multidimensional data is measured using the percentage of runs ($perc$) in 100 runs when estimate of global minimum differs from f^*_{min} by less than 10^{-4}. f^*_{min} has been estimated independently running the algorithm relatively longer.

3.1. *Data Sets*

Several sets of multidimensional points corresponding to well understood geometric objects are needed for the experimental investigation. We want to choose difficult test problems, i.e. difficult to visualize geometric objects. The data with desired properties correspond to the multidimensional objects equally extending in all dimensions of the original space, e.g. sets of vertexes of multidimensional cubes and simplexes. Dissimilarity between vertexes is measured by the distance in the original vector space defined by its metric; we consider Euclidean and city block metrics. Global optimization problems of different difficulty can be constructed by defining dimensionality of the original spaces. Below we use shorthand "cube" and "simplex" for sets of their vertexes.

The number of vertexes of multidimensional cube is $n = 2^{\text{dim}}$, and the dimensionality of global minimization problem is $N = 2^{\text{dim}+1}$. The coordinates of i-th vertex of a dim-dimensional cube are equal either to 0 or to 1, and they are defined by binary code of $i = 1, ..., n$. Vertexes of multidimensional simplex can be defined by

$$v_{ij} = \begin{cases} 1, \text{ if } i = j+1, \\ 0, \text{ otherwise}, \end{cases} i = 1, \ldots, \dim + 1, j = 1, \ldots, \dim. \quad (5)$$

Dimensionality of this global minimization problem is $N = 2 \times (\dim + 1)$.

3.2. *Impact of Parameters of the Algorithm to Performance*

Performance of the algorithm depend on the values of the parameters p and N_{init}. The impact of the parameters of the algorithm to reliability of the algorithm has been evaluated experimentally visualizing geometrical data sets of different dimensionality. t_c=10s has been used. Performance (*perc*) of the algorithm with Euclidean metric visualizing multidimensional cubes with different values of the parameters p and N_{init} set is shown in Table 1. The highest estimated values of *perc* for each problem are indicated. The value of the parameter p does not give much influence to the performance of the algorithm with Euclidean metric. N_{init}=2000 seems most favorable. For larger values of N_{init}, initialization takes longer and decreases the overall performance of the algorithm. It is especially exposed for larger problems. The balance between initialization and genetic algorithm is required.

Performance (*perc*) of the algorithm with city block metric visualizing multidimensional data with different values of the parameters p and N_{init} set is shown in Table 2. The highest estimated values of *perc* for each problem are indicated. There is no clear tendency of the influence of value of the parameters to the performance of the algorithm with city block metric. However p=60 and N_{init}=6000 seem most favorable values of the parameters.

3.3. *Investigation of the Parallel Algorithm*

The parallel version of the hybrid algorithm has been used to visualize multidimensional data on Sun Fire E15k high performance computer at EPCC. p=60, N_{init}=2000, t_c=10s have been used for the algorithm with Euclidean distances. p=60, N_{init}=6000, t_c=10s have been used for the algorithm with city block distances. For the assessment of the performance, the results of visualization using the hybrid algorithm with city block metric are presented in Table 3. Columns represent results with different numbers

Table 1. Performance (*perc*) of the algorithm with Euclidean metric.

	p=20	p=40	p=60	p=80	p=100
		6-dimensional cube, f^*_{min} =0.3505			
N_{init}=1000	**58**	**58**	**58**	**58**	**57**
N_{init}=2000	**57**	**57**	**57**	**57**	**57**
N_{init}=4000	50	50	50	51	51
N_{init}=6000	50	50	51	50	50
N_{init}=8000	52	52	52	52	52
N_{init}=10000	46	46	46	45	46
N_{init}=12000	41	41	41	41	38
		7-dimensional cube, f^*_{min} =0.3629			
N_{init}=1000	46	46	46	46	46
N_{init}=2000	**53**	**53**	**53**	**53**	**53**
N_{init}=4000	**50**	**50**	**50**	**50**	**50**
N_{init}=6000	45	44	45	45	45
N_{init}=8000	42	42	42	42	42
N_{init}=10000	43	43	43	43	43
N_{init}=12000	39	39	36	39	40
		8-dimensional cube, f^*_{min} =0.3715			
N_{init}=1000	**52**	**52**	**52**	**52**	**52**
N_{init}=2000	**54**	**54**	**54**	**54**	**54**
N_{init}=4000	47	47	47	47	47
N_{init}=6000	2	1	1	4	3
N_{init}=8000	0	0	0	2	2
N_{init}=10000	0	0	0	0	0
N_{init}=12000	0	0	0	0	0

of processors used. Performance improvement is significant for all considered problems comparing with the performance on single processor. When 8 parallel processors are used, reliability of visualization of 5-dimensional cube is increased 4 times, and reliability of visualization of 16-dimensional simplex is increased 7 times. Parallelization has increased dimensionality of reliably visualized simplexes from 12 to 14.

The results of similar experiment on Vytautas Magnus University (VDU) cluster of personal computers are presented in Tables 4 and 5. Columns represent results with different numbers of processes used. When 8 processes are used, reliability of visualization of 5-dimensional cube is increased 4 times, and reliability of visualization of 18-dimensional simplex is increased 8 times. Parallelization has increased dimensionality of reliably visualized simplexes from 12 to 16. However increased number of processes on VDU cluster does not necessary increase reliability of the algorithm.

As the same estimated values f^*_{min} has been used for VDU cluster and Sun Fire E15k high performance computer, results let compare performance of the algorithm on both computer systems. Results of experiments show

Table 2. Performance (*perc*) of the algorithm with city block metric.

	p=20	p=40	p=60	p=80	p=100
	5-dimensional cube, f^*_{min} =0.3313				
N_{init}=1000	23	36	**40**	34	28
N_{init}=2000	**41**	**39**	**38**	26	26
N_{init}=4000	30	24	28	33	32
N_{init}=6000	27	25	**39**	**38**	**39**
N_{init}=8000	34	26	33	33	**37**
N_{init}=10000	30	**36**	33	**36**	**37**
N_{init}=12000	10	31	4	9	**38**
	15-dimensional simplex, f^*_{min} =0.3439				
N_{init}=2000	58	68	**93**	83	87
N_{init}=4000	57	83	86	89	85
N_{init}=6000	55	85	**93**	89	85
N_{init}=8000	52	80	88	**92**	87
N_{init}=10000	56	87	89	**97**	**96**
N_{init}=12000	43	84	91	88	83
	16-dimensional simplex, f^*_{min} =0.3484				
N_{init}=2000	29	57	71	**72**	66
N_{init}=4000	24	65	68	**72**	63
N_{init}=6000	28	56	**72**	61	53
N_{init}=8000	27	56	**76**	71	58
N_{init}=10000	27	64	**74**	71	48
N_{init}=12000	27	58	**78**	69	50
N_{init}=14000	28	53	65	60	65
	17-dimensional simplex, f^*_{min} =0.3526				
N_{init}=2000	26	**37**	34	22	19
N_{init}=4000	18	35	**36**	26	15
N_{init}=6000	21	**41**	**39**	22	18
N_{init}=8000	24	31	28	25	14
N_{init}=10000	24	32	**41**	20	7
N_{init}=12000	17	**36**	29	21	11

that reliability of the algorithm on VDU cluster is higher than on Sun Fire E15k high performance computer. This is most probably because nodes on VDU cluster are more powerful than nodes on the used Sun Fire E15k.

The results of similar experiment on Institute of Physiology and Rehabilitation c/o Kaunas University of Medicine (KMU PRI) cluster of personal computers are presented in Table 6. Columns represent results with different numbers of processes used. When 4 processes are used, reliability of visualization of 5-dimensional cube is increased 2.5 times.

4. Conclusions

The value of the parameter p does not give much influence to the performance of the hybrid visualization algorithm with Euclidean metric.

Table 3. Performance of the parallel algorithm with city block metric on SUN Fire E15k.

dim	1	2	4	6	8
			cubes		
3	100	100	100	100	100
4	96	100	100	100	100
5	14	22	28	33	55
			simplexes		
5	100	100	100	100	100
6	100	100	100	100	100
7	100	100	100	100	100
8	100	100	100	100	100
9	100	100	100	100	100
10	100	100	100	100	100
11	100	100	100	100	100
12	100	100	100	100	100
13	93	100	100	100	100
14	70	90	100	100	100
15	25	45	77	81	94
16	8	17	34	43	56

Table 4. Performance of the parallel algorithm with Euclidean metric on VDU personal computing cluster.

dim	1	2	3	4	5	6	7	8
				cubes				
5	100	100	100	100	100	100	100	100
6	40	76	94	96	98	99	99	100
7	53	73	78	89	96	98	100	100
8	54	77	94	94	96	98	98	99

N_{init}=2000 seems most favorable value for the algorithm with Euclidean metric. p=60 and N_{init}=6000 are most favorable values of the parameters for the algorithm with city block metric.

Performance improvement using parallel version of the algorithm is significant for all considered problems comparing with the performance on single processor. Efficiency of parallelization of the algorithm could be investigated by comparing results when total computational time is the same for any configuration of parallel computer. This would be the most valuable direction for the future analysis of the implemented parallel algorithm.

Acknowledgments

The research is supported by Lithuanian State Science and Studies Foundation, the NATO Reintegration grant CBP.EAP.RIG.981300, and the HPC-Europa programme, funded under the European Commission's Research

Table 5. Performance of the parallel algorithm with city block metric on VDU personal computing cluster.

dim	1	2	3	4	5	6	7	8
				cubes				
3	100	100	100	100	100	100	100	100
4	100	100	100	100	100	100	100	100
5	24	49	73	86	94	90	95	96
6	6	7	8	16	19	31	28	31
				simplexes				
3	100	100	100	100	100	100	100	100
4	100	100	100	100	100	100	100	100
5	100	100	100	100	100	100	100	100
6	100	100	100	100	100	100	100	100
7	100	100	100	100	100	100	100	100
8	100	100	100	100	100	100	100	100
9	100	100	100	100	100	100	100	100
10	100	100	100	100	100	100	100	100
11	100	100	100	100	100	100	100	100
12	100	100	100	100	100	100	100	100
13	99	100	100	100	100	100	100	100
14	98	100	100	100	100	100	100	100
15	73	99	100	100	100	100	100	100
16	37	93	93	99	100	100	100	100
17	29	32	69	84	92	99	92	97
18	10	29	79	76	74	78	77	80
19	0	1	9	13	18	31	29	21
20	0	2	5	32	23	21	21	13

Table 6. Performance of the parallel algorithm with city block metric on KMU PRI personal computing cluster.

dim	1	2	3	4
		cubes		
3	100	100	100	100
4	100	100	100	100
5	34	73	79	83
6	4	14	13	14

Infrastructures activity of the Structuring the European Research Area programme, contract number RII3-CT-2003-506079.

References

1. I. Borg and P. Groenen, *Modern Multidimensional Scaling*, Springer, New York (1997).
2. T. Cox and M. Cox, *Multidimensional Scaling*, Chapman and Hall/CRC, Boca Raton (2001).
3. J. De Leeuw and W. Heiser, Theory of multidimensional scaling, In: P. R. Kr-

ishnaiah (ed.), *Handbook of Statistics, v. 2*, North Holland, Amsterdam, pp. 285–316 (1982).

4. P. Groenen, R. Mathar and W. Heiser, The majorization approach to multidimensional scaling for minkowski distances, *Journal of Classification* **12**, 3–19 (1995).
5. P. Groenen, R. Mathar and J. Trejos, Global optimization methods for MDS applied to mobile communications, In: W. Gaul, O. Opitz and M. Schander (eds.), *Data Analysis: Scientific Models and Practical Applications*, Springer, 459–475 (2000).
6. R. Mathar, A hybrid global optimization algorithm for multidimensional scaling, In: R. Klar and O. Opitz (eds.), *Classification and Knowledge Organization*, Springer, Berlin, 63–71 (1996).
7. R. Mathar and A. Žilinskas, On global optimization in two-dimensional scaling, *Acta Applicandae Mathematicae* **33**, 109–118 (1993).
8. W. Press *et al.*, *Numerical Recipes in C++*, Cambridge University Press, Cambridge (2002).
9. A. Žilinskas and J. Žilinskas, Two level minimization in multidimensional scaling, *Journal of Global Optimization*, submitted.
10. A. Žilinskas and J. Žilinskas, Parallel hybrid algorithm for global optimization of problems cccurring in MDS based visualization, *Computers & Mathematics with Applications*, accepted.
11. E. Cantú-Paz, *Efficient and Accurate Parallel Genetic Algorithms*, Kluwer Academic Publishers (2000).
12. Message Passing Interface Forum, *MPI: A Message-Passing Interface Standard (version 1.1)*, Technical report (1995).

MULTIDIMENSIONAL SCALING IN PROTEIN AND PHARMACOLOGICAL SCIENCES

J. ŽILINSKAS

Institute of Mathematics and Informatics, Akademijos 4
Vilnius LT-08663, Lithuania, E-mail: julius.zilinskas@mii.lt

Multidimensional scaling is a technique for visualization of multidimensional data. In this paper pharmacological data are visualized using multidimensional scaling for the visual analysis of properties of adrenoceptors (which form a class of G-protein-coupled receptors) and ligands. The aim of visualization is to provide useful information for prediction of structural features of proteins and drug design. For the implementation of a multidimensional scaling technique a difficult global optimization problem should be solved. To attack such a problem a hybrid global optimization method is used, where an evolutionary global search is combined with a local descent.

1. Introduction

A protein is a complex organic compound that consists of amino acids joined by peptide bonds; the sequence of amino acids is coded by the DNA. Proteins are building blocks of all living cells and viruses; they play structural roles or, upon binding of a ligand (small molecule that binds to a protein), serve for signaling, transport and catalytic processes. The arrangements of the amino acids in the three-dimensional space into a stable, low energy, structure is key to protein function. The protein structure prediction is a fundamental scientific problem and it is regarded as a holy grail in computational chemistry, molecular and structural biology.[1]

The measured binding affinity between proteins and ligands may provide useful information for prediction of structural features of proteins and drug design. The binding affinity data is usually represented through a matrix, one dimension formed by the different ligands tested in a series of experiments while the other dimension represents the different proteins. Visual analysis of the matrix of measured values is difficult. Ruuskanen *et al.*[2] used principal component analysis and binary trees of binding affinity data to analyze clustering of the receptors and ligands. Binary trees provide an artificial clustering by pairs that can be very misleading, but on the other hand it is easiest for the human mind to interpret. Usually some data are not represented when principal components are visualized. Although three most significant axes were visualized by Ruuskanen *et al.*[2] about 15% and 19% of data were not represented.

In this paper pharmacological data of binding affinity between adrenoceptors (cell surface messenger proteins) and ligands (natural neurotransmitters or pharmacological drugs) have been visualized using multidimensional scaling (MDS) – an exploratory technique for analysis of multidimensional data.[3,4] The points $\mathbf{x}_i = (x_{i1}, \ldots, x_{im}), i = 1, \ldots, n$ representing n objects in m-dimensional embedding space should be found fitting pairwise distances of points to given pairwise dissimilarities of the objects (δ_{ij}). The two-dimensional embedding space ($m = 2$) is of special interest, when MDS is used to visualize multidimensional data.

The implementation of a MDS method is reduced to minimization of a fitness criterion, e.g. the so called *STRESS* function:

$$S(X) = \sum_{i<j}^{n} (d_{ij}(X) - \delta_{ij})^2, \tag{1}$$

where $X = ((x_{11}, \ldots, x_{n1}), \ldots, (x_{1m}, \ldots, x_{nm}))$; $d_{ij}(X)$ denotes the distance between the points $\mathbf{x}_i$ and $\mathbf{x}_j$. In this paper Euclidean and city block distances are used. MDS is a difficult global optimization problem. Although *STRESS* is defined by an analytical formula, which seems rather simple, its minimization is difficult.

In the case of MDS with Euclidean distances an algorithm combining local descent and evolutionary search has been proposed by Mathar and Žilinskas.[5] Such an algorithm is shown to be most reliable of known MDS algorithms experimentally tested.[6,7] Similar hybrid global optimization algorithm for multidimensional scaling with Euclidean and city block distances[8] is used in this paper. The details of the current version of the algorithm can be found in the previous chapter.[9] The values of parameters for algorithm used in this paper are: the size of population $p = 60$, the number of random points generated during initialization $N_{init} = 6000$, predetermined computing time $t_c = 100s$. The images of multidimensional data with smallest *STRESS* are shown in the figures in this paper. Besides of qualitative assessment of informativeness of the images it is interesting to compare "visualization errors" quantitatively. To exclude the impact of number of objects and of used metric a relative error

$$f(X) = \sqrt{\frac{S(X)}{\sum_{i<j} \delta_{ij}^2}}, \tag{2}$$

is used for comparison.

2. Multidimensional Scaling of Pharmacological Binding Affinity Data

Adrenergic receptors (adrenoceptors) are a class of G-protein coupled receptors. The adrenoceptors are cell surface messenger proteins that mediate the physiological effects of the hormones/neurotransmitters adrenaline and noradrenaline, and are target molecules of several clinically important drugs. The adrenoceptors can be divided into three main classes, α_1-, α_2- and β-adrenoceptors that have different physiological effects. Each of these classes is divided into subtypes which are distinct receptors encoded by different genes, have quite similar but not identical amino acid sequences, are expressed in different tissues, have different mode of regulation and their own ligand binding specificity.

Pharmacological data, e.g. K_i, inhibition constants values, are constants which under given experimental conditions link the affinity of a given ligand for a given receptor protein. Inhibition constants are obtained by competition binding assays, which test the ability of ligand to displace a radioactive ligand (radioligand) from the binding site. The K_i is an equilibrium dissociation constant, defined as the concentration of competing ligand which would bind to half the receptor binding sites at equilibrium in the absence of radioligand or other competitors. A ligand can be a natural neurotransmitter or pharmacological drug that binds to a receptor, an agonist drug activates the receptor upon binding while an antagonist drug blocks the action of the receptor. Pharmacological data is usually represented through a matrix, one dimension formed by the different ligands tested in a series of experiments while the other dimension represents the different receptors used, which can be from different types or subtypes, or from different species, or engineered mutants of these.

Analysis of such pharmacological data is very important. A correlation between structural features of a group of ligands and similar variations in binding affinity across subtypes and species may provide useful information for drug design. In the other hand, analysis of the pharmacological properties of the adrenoceptors can help to predict structural features of them.

Multidimensional scaling of pharmacological binding affinity data visualizing properties of adrenoceptors and properties of ligands can be used for visual representation of pharmacological data and further visual analysis. Hybrid global optimization algorithm with Euclidean and city block metrics has been used to visualize binding affinity data from different studies in this paper. Dissimilarities of receptors have been calculated using pairwise Euclidean and city block distances between vectors of the $\log_{10}$-transformed binding affinities representing properties of the receptors. Dissimilarities of ligands have been calculated using distances between vectors representing ligands.

3. Results of Experiments

3.1. *Visualization of Properties of Three Human and Five Zebrafish α_2-Adrenoceptors and 20 Ligands*

Structural, pharmacological and functional properties of three human and five zebrafish α_2-adrenoceptors have been analyzed by Ruuskanen *et al.*[2] The ligand binding affinities of them have been tested with respect to 20 ligands known to bind to the human α_2-adrenoceptors. In this paper the hybrid global optimization algorithm with Euclidean and city block metrics has been used to visualize binding affinity data of the same eight adrenoceptors and the 20 ligands.

Images of the binding affinity data as properties of three human and five zebrafish α_2-adrenoceptors are shown in Fig. 1. Human receptors are indicated by the first letter "h" and zebra fish receptors by "z". Images in Fig. 1 show close clustering of $z\alpha_{2Da}$ and $z\alpha_{2Db}$. There is a possible clustering of zebrafish receptors in the centers of images and human receptors around them. Adrenoceptors $h\alpha_{2B}$ and $z\alpha_{2Da}$ are visualized closer to each other than $h\alpha_{2A}$ to $z\alpha_{2A}$, $h\alpha_{2B}$ to $z\alpha_{2B}$, and $h\alpha_{2C}$ to $z\alpha_{2C}$. It is especially exposed in the image acquired by multidimensional scaling with city block metric.

Images of the binding affinity data as properties of the 20 ligands are shown in Fig. 2. The ligands are numbered according to Ruuskanen *et al.*[2] Agonists are indicated by the symbol "+". Antagonists are indicated by the symbol "x". Images in Fig. 2 show clustering of agonists and antagonists. However, agonists 2 and 3 are near cluster of antagonists.

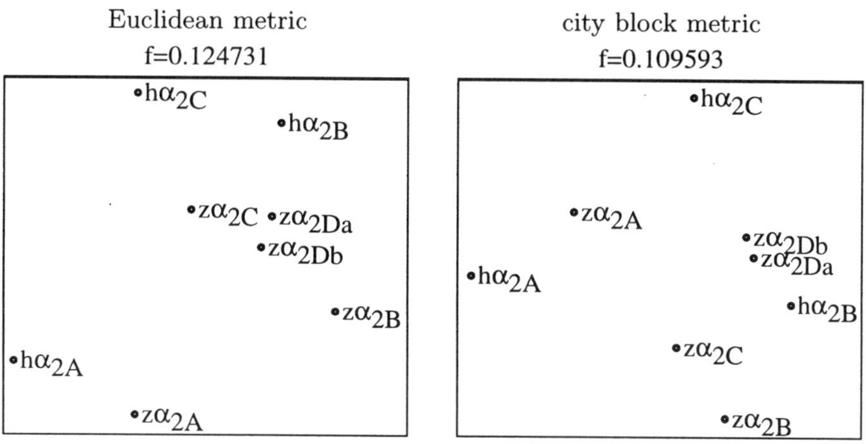

Figure 1. Images of the properties of human and zebrafish α_2-adrenoceptors.

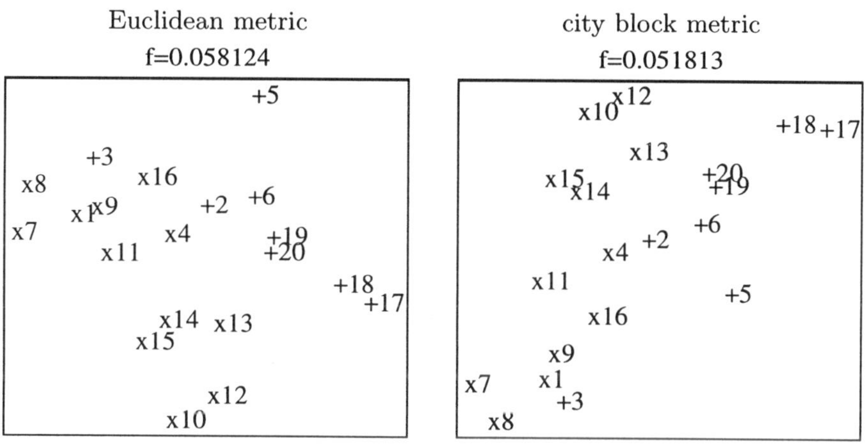

Figure 2. Images of properties of 20 ligands binding human and zebrafish α_2-adrenoceptors.

3.2. *Visualization of Properties of Human, Rat, Guinea Pig and Pig* α_2*-Adrenoceptors*

Binding properties of five radioligands to human, rat, guinea pig and pig α_2-adrenoceptors have been analyzed by Uhlen *et al.*[10] In this paper the hybrid global optimization algorithm with Euclidean and city block metrics has been used to visualize binding affinity data as properties of these adrenoceptors. Images of the binding affinity data as properties of human,

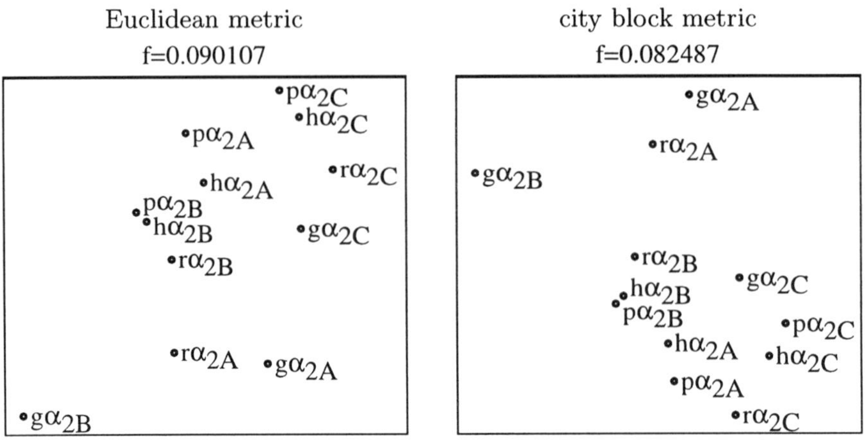

Figure 3. Images of the properties of human, rat, guinea pig and pig α_2-adrenoceptors.

rat, guinea pig and pig α_2-adrenoceptors are shown in Fig. 3. Human receptors are indicated by the first letter "h", rat receptors by "r", guinea pig receptors by "g" and pig receptors by "p".

Images in Fig. 3 show close clustering of hα_{2B}, pα_{2B} and rα_{2B}. Properties of guinea pig α_{2B}-adrenoceptor are quite different to those of other visualized adrenoceptors. α_{2A}-adrenoceptors of different species form two clusters: hα_{2A} with pα_{2A} and rα_{2A} with gα_{2A}. α_{2C}-adrenoceptors of different species form a cluster. Images in Fig. 3 show that properties of human and pig α_2-adrenoceptors are quite similar comparing with properties of rat and guinea pig adrenoceptors.

3.3. *Visualization of Properties of Wild Type and Mutant α_1-Adrenoceptors and Ligands*

Critical amino acids of α_{1a}- and α_{1b}-adrenoceptors have been analyzed by Hwa *et al.*[11] The goal of the analysis is to understand which amino acids are involved in the specificity for binding certain ligands. Therefore binding affinities of α_{1a}- and α_{1b}-adrenoceptors and their engineered mutants with one or two amino acids changed have been compared. For this reason mutants have been engineered changing one or two amino acids in α_{1b}- to their equivalent in α_{1a}-adrenoceptor. Binding affinities of the mutants have been analyzed searching for the change becoming alike to α_{1a}-adrenoceptor. Similarly mutants of α_{1a}-adrenoceptor have been engineered. In this paper the hybrid global optimization algorithm with Euclidean and city block

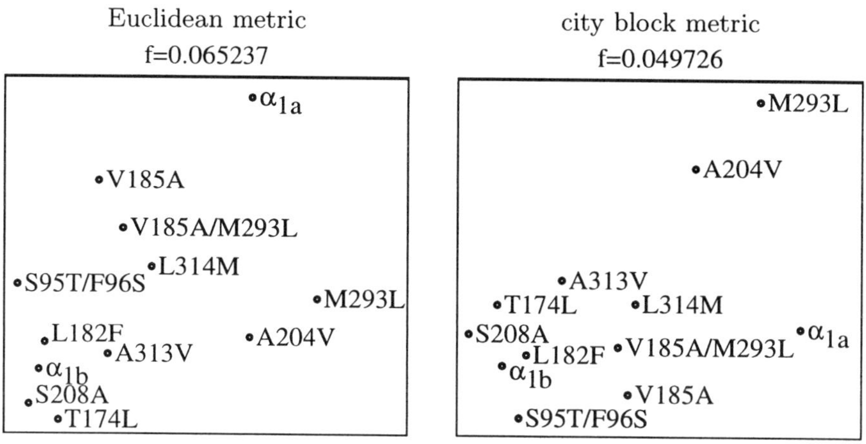

Figure 4. Images of the properties of wild type and mutant α_1-adrenoceptors.

metrics has been used to visualize binding affinity data of wild type and mutant α_1-adrenoceptors and ligands.

Images of the binding affinity data as properties of the receptors are shown in Fig. 4. The images show that properties of the mutants T174L, L182F and S208A of α_{1b}-adrenoceptor are similar to properties of the wild type. Mutations S95T/F96S and A204V of α_{1b}-adrenoceptor change binding properties significantly, but do not make them alike to properties of α_{1a}-adrenoceptor. Similarly mutation M293L of α_{1a}-adrenoceptor change binding properties significantly, but do not make them alike to properties of α_{1b}-adrenoceptor. Properties of mutant L314M of α_{1b}-adrenoceptor as well as of mutants V185A and V185A/M293L of α_{1a}-adrenoceptor are similar and are visualized in the middle between properties of wild type α_{1a}- and α_{1b}-adrenoceptors.

Images of binding affinity data as properties of the ligands are shown in Fig. 5. Agonists are indicated by the symbol "+". Antagonists are indicated by the symbol "x". Images show clustering of agonists and antagonists. However images show that the properties of agonist "3" (Methoxamine) are quite different from other ligands. Image acquired using multidimensional scaling with city block distances shows that properties of antagonist "8" (5 Methylurapidil) are more similar to properties of agonists than to those of other antagonists, but image acquired using multidimensional scaling with Euclidean distances does not show this property.

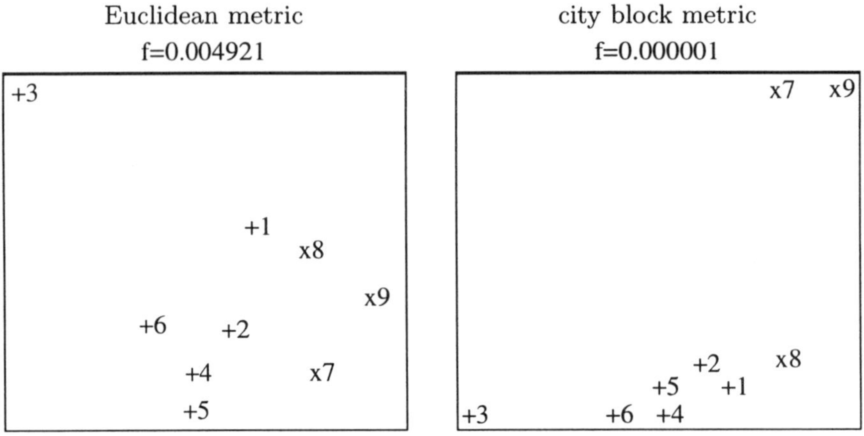

Figure 5. Images of the properties of 9 ligands binding wild type α_{1a}- and wild type and mutant α_{1b}-adrenoceptors.

3.4. *Performance of Algorithms for Multidimensional Scaling*

Images of the pharmacological data acquired using multidimensional scaling with Euclidean and city block metric show similar properties of adrenoceptors or ligands. However some properties are especially exposed in the images of the pharmacological data shown as properties of receptors acquired by multidimensional scaling with city block metric. The relative errors are similar as well, however they are smaller when city block metric is used. The relative errors are smaller when the pharmacological data is visualized as properties of ligands.

Performance of the hybrid global optimization algorithm for multidimensional scaling in visualization of multidimensional data is shown in Table 1. Minimal, average and maximal estimates of global minimum in 100 runs (f^*_{min}, f^*_{mean} and f^*_{max}) are presented in the table to show quality of the found solutions. The percentage of runs (*perc*) when estimate of global minimum differs from f^*_{min} by less than 10^{-4} is presented in the table as a criterion of reliability of the algorithm. Estimates of global minimum in 100 runs are very similar and differ by less than 10^{-4} in most cases. However they differ by less when Euclidean metric is used, what suggests that problems with city block metric are more difficult to solve.

Table 1. Performance of the hybrid global optimization algorithm for multidimensional scaling.

Euclidean metric				city block metric			
f^*_{min}	f^*_{mean}	f^*_{max}	*perc*	f^*_{min}	f^*_{mean}	f^*_{max}	*perc*
Results presented in Fig. 1							
0.1247	0.1247	0.1247	100	0.1096	0.1096	0.1096	100
Results presented in Fig. 2							
0.0581	0.0581	0.0581	100	0.0518	0.0518	0.0518	100
Results presented in Fig. 3							
0.0901	0.0901	0.0901	100	0.0825	0.0825	0.0826	78
Results presented in Fig. 4							
0.0652	0.0652	0.0652	100	0.0497	0.0497	0.0497	100
Results presented in Fig. 5							
0.0049	0.0049	0.0049	100	1.323e-06	1.337e-06	1.348e-06	100

4. Conclusions

Multidimensional scaling can be used in protein science for simplification and visual analysis of complex pharmacological data. Images of the binding affinity data acquired using multidimensional scaling with Euclidean and city block metric show similar properties. However some properties are especially exposed when city block metric is used. Moreover the relative errors are smaller when city block metric is used. However problems with city block metric are more difficult to solve.

This is to my knowledge the first report of visualization of the complex effect of point mutants on the pharmacological profile of wild type and mutated receptors in a simple view. The pharmacologist community would benefit from using multidimensional scaling for visualization of pharmacological data.

Acknowledgments

The research is supported by Lithuanian State Science and Studies Foundation and the NATO Reintegration grant CBP.EAP.RIG.981300. I would like to thank H. Xhaard for suggestion of the case studies and helpful discussions.

References

1. C. A. Floudas, Research challenges, opportunities and synergism in systems engineering and computational biology, *AIChE Journal* **51**(7), 1872–1884 (2005).
2. J. O. Ruuskanen, J. Laurila, H. Xhaard, V.-V. Rantanen, K. Vuoriluoto, S. Wurster, A. Marjamaki, M. Vainio, M. S. Johnson and M. Scheinin, Conserved structural, pharmacological and functional properties among the three human and five zebrafish α_2-adrenoceptors, *British Journal of Pharmacology* **144**(2), 165–177 (2005).
3. I. Borg and P. Groenen, *Modern Multidimensional Scaling*, Springer, New York (1997).
4. T. Cox and M. Cox, *Multidimensional Scaling*, Chapman and Hall/CRC, Boca Raton (2001).
5. R. Mathar and A. Žilinskas, On global optimization in two-dimensional scaling, *Acta Applicandae Mathematicae* **33**, 109–118 (1993).
6. P. Groenen, R. Mathar and J. Trejos, Global optimization methods for MDS applied to mobile communications, In: W. Gaul, O. Opitz and M. Schander (eds.), *Data Analysis: Scientific Models and Practical Applications*, Springer, 459–475 (2000).
7. R. Mathar, A hybrid global optimization algorithm for multidimensional scaling, In: R. Klar and O. Opitz (eds.), *Classification and Knowledge Organization*, Springer, Berlin, 63–71 (1996).
8. A. Žilinskas and J. Žilinskas, Parallel hybrid algorithm for global optimization of problems cccurring in MDS based visualization, *Computers & Mathematics with Applications*, accepted.
9. A. Varoneckas, A. Žilinskas and J. Žilinskas, Multidimensional scaling using parallel genetic algorithm, *in the same volume* (2006).
10. S. Uhlén, M. Dambrova, J. Näsman, H. B. Schiöth, Y. Gu, A. Wikberg-Matsson and J. E. S. Wikberg, [^{3}H]RS79948-197 binding to human, rat, guinea pig and pig α_{2A}-, α_{2B}- and α_{2C}-adrenoceptors. Comparison with MK912, RX821002, rauwolscine and yohimbine, *European Journal of Pharmacology* **343**(1), 93–101 (1998).
11. J. Hwa, R. M. Graham and D.M. Perez, Identification of critical determinants of α_1-adrenergic receptor subtype selective agonist binding, *Journal of Biological Chemistry* **270**(39), 23189–23195 (1995).

ON DISSIMILARITY MEASUREMENT IN VISUALIZATION OF MULTIDIMENSIONAL DATA

A. ŽILINSKAS
Institute of Mathematics and Informatics, VMU, Akademijos str. 4
Vilnius, 08663, Lithuania

A. PODLIPSKYTĖ
Institute of Psychophysiology and Rehabilitation Vyduno str. 4
Palanga, 00135, Lithuania

Multidimensional scaling (MDS) is a prospective technique to the visualization and exploratory analysis of multidimensional data. By means of MDS algorithms a two dimensional representation of a set of points in a high dimensional (original) space can be obtained, where distances between the points in the two dimensional embedding space represent dissimilarity of multidimensional points. The latter normally is measured by the Euclidean distance, although the alternative measures can be advantageous. In the present paper we investigate influence of the choice of dissimilarity measure (distances in the original space) to the visualization results.

1. Introduction

To visualize multidimensional data, two or three-dimensional objects representing the data should be created.[1,2] We aim to represent sets of multidimensional vectors (original data) by sets of two-dimensional points (image). A mapping from a multidimensional space to a two dimensional space should ensure some similarity between the structure of an original data and of its image. For example, a multidimensional vector can represent an individual from a sample of patients suffering from a disease. We want to obtain a two-dimensional set of points where images of patients with similar diagnoses form distinguishable clusters.

Dissimilarity between elements in a vector space can be measured by a norm of their difference, i.e. by a distance between the corresponding points. A mapping is considered precisely preserving the structure of a data set if the distances between the original points and the distances between their images are equal. Practically we want to minimize an error of approximation of the distances in the original space by the distances in the embedding space. For example, by the classical method of principal component analysis a multidimensional space is mapped to a space of lower dimensionality linearly with the minimal square error.[3] In the class of nonlinear mappings the methods with smaller approximation error can be found. For the background, algorithmic implementation, and applications of the most popular methods we refer to literature.[4-7]

2. Basics of Multidimensional Scaling

Let $X_i \in R^n$, $i = 1,..,k$ be the data intended to visualize. We are searching for a set of two dimensional points $Y_i \in R^2$, $i = 1,..,k$ whose interpoint distances $d_{ij}(Y)$ well approximate interpoint distances $\delta_{ij} = \|X_i - X_j\|$. The approximation error normally is defined as

$$stress = \sum_{i=1}^{k} \sum_{j=i+1}^{k} w_{ij} \left(d_{ij}(Y) - \delta_{ij}\right)^2, \tag{1}$$

$$sstress = \sum_{i=1}^{k} \sum_{j=i+1}^{k} w_{ij} \left(d_{ij}^{\,2}(Y) - \delta^2_{ij}\right)^2 \tag{2}$$

where $w_{ij} \geq 0$ are weights. The image is a set of two-dimensional vectors Y_i, $i = 1,..,k$ corresponding to minimum point of (1) or (2). A crucial problem in the implementation of MDS methods is minimization of an approximation error. The problem is especially difficult because of multimodality of these criteria. A suitable method should be chosen from known global optimization methods,[8] or a special method should be tailored taking into account properties of the considered criteria.[9,10] The properties of minimization problems obviously depend on the norm in R^2 used to calculate $d_{ij}(Y)$. For example, the usage of the general Minkowski metric versus the Euclidean metric implies additional minimization difficulties caused by nondifferentiability of the objective function. The influence of the metric used to calculate δ_{ij} is not explicit, e.g. it does not influence smoothness of the objective function. However, it can be crucial in defining qualitative properties of the objective function generally assessed as optimization difficulties.

It is well known that for the conceptually similar sets of points in the spaces of different dimensionality the structures of Euclidean distances can be different; see[11] for an example of random uniformly distributed points. The use of different norms in original and embedding spaces can help to bring both distance structures closer. We will use Euclidean metric in the embedding space R^2 since a two dimensional image is supposed for heuristic analysis by humans, and humans are used to measure distances in Euclidean metric. Several widely used norms are applied to calculate δ_{ij}.

3. Dissimilarities in the Original Space

A distance in the multidimensional original space can be considered as a measure of dissimilarity, and it is defined by a corresponding norm. The most widely used norm is the Euclidean norm. The Minkowski norm is a generalization of the Euclidean norm; the distance between X_i and X_j is defined by the formula

$$\delta_{ij} = \sqrt[p]{\sum_{r=1}^{n} (x_{ir} - x_{jr})^p}, \quad X_i = (x_{i1}, \dots, x_{in}). \tag{3}$$

Special cases of Minkowski distance are: the maximum distance ($p = \infty$), $\delta_{ij} = \max_{r=1,\dots,n} |x_{ir} - x_{jr}|$, and city block distance ($p = 1$), $\delta_{ij} = \sum_{r=1}^{n} |x_{ir} - x_{jr}|$

We will apply in our investigation also the Mahalanobis distance

$$\delta_{ij} = \left((X_i - X_j)^T \Sigma^{-1} (X_i - X_j)\right)^{1/2} \tag{4}$$

where $\Sigma = \overline{X} \cdot \overline{X}^T, \quad \overline{X} = X - M = (X_1 - M | X_2 - M | \dots | X_k - M)$

$M = (m_1, \dots, m_n)^T, m_r = \frac{1}{k} \sum_{i=1}^{k} x_{ir}$.

The dissimilarity will also be measured using correlation between multidimensional vectors

$$\delta_{ij} = 1 - \frac{\sum_{r=1}^{n} (x_{ir} - \overline{x}_i)(x_{jr} - \overline{x}_j)}{\left(\sum_{r=1}^{n} (x_{ir} - \overline{x}_i)^2 \sum_{r=1}^{n} (x_{jr} - \overline{x}_j)^2\right)^{1/2}}, \quad \overline{x}_i = \frac{1}{n} \sum_{r=1}^{n} x_{ir}. \tag{5}$$

4. Data Sets

The testing starts with an artificial data set of 64 points in the six dimensional space constituting vertices of the unit hypercube. All six dimensions for this data set are equally important, and any projection to two dimensions corrupts the distances between vertices. However, the geometric structure of this data set is absolutely clear. A good visualization method should expose various symmetries of the original data and its substructures formed by 2^i points.

A data set used by many researchers in analysis of structures and pattern recognition is Iris flower data proposed in classical paper by R. Fisher.[12] The data set consists of four measurements from 150 flowers: 50 Iris-setosa, 50 Iris-versicolor and 50 Iris-virginica. It was reduced to 80 items because of limitations of the MDS codes in the statistical packages STATISTICA and SPSS. The measurements of the data are length and width of sepal and petal leaves. Different statistical methods have been applied to analyze the data. We are interested to distinguish the species considering vectors of four variables. It is well established that one cluster of points in the four dimensional space is well separable from the other containing data of two species. Two other clusters can be separated but not very precisely.

Two sets of real world data is taken from the data basis of the Institute of Psychophysiology and Rehabilitation (Kaunas University of Medicine). Both data sets contain 80 seven dimensional vectors. The modest size of data sets is chosen because of limitations mentioned above. The first data set contains the data of 80 patients recorded by means of polysomnographic technique during a night. Such a data presents objectively measured sleep criteria of the considered patients.[13] The second data set is formed on the basis of answers of the patients to the questions of the questionnaire of Pittsburg Sleep Quality Index.[14] In the next sections these data sets are referenced as "objective sleep assessment" and "subjective sleep assessment" correspondingly. Both data sets have been analyzed by means of MDS technique using the Euclidean distance as a dissimilarity measure between the original vectors.[15] The images of these data sets have exposed three distinguishable clusters. The multidimensional data of the clusters have been analyzed by standard statistical techniques. The results of statistical analysis have shown that the clusters correspond to three subsets of patients: with good sleep quality, suffering from some sleep disorders, and with serious illness.

5. Methods

Several standard algorithms were considered together with algorithm developed by the authors. The MDS algorithms from the packages STATISTICA, SPSS, NCSS were tested. The manuals of these packages present brief information on the included algorithms and their implementations. The algorithm in STATISTICA is based on minimization of *stress* criterion, while SPSS is based on minimization of *sstress* criterion. Two algorithms from the package NCSS were tested; both of them are based on minimization of *stress* criterion but the first algorithm is metric and the second is non-metric. No information on the applied minimization methods and on the weights in formulas (1), (2) is presented there.

A MDS algorithm based on hybrid optimization of (1) and (2) combining evolutionary global search and local descent[9] has been developed by the authors.[15] In the subsequent sections, this algorithm is called hybrid MDS *stress* and hybrid MDS *sstress* indicating the optimization criterion: (1) or (2). The weights in the latter formulae are chosen equal to all summands and defined as correspondingly:

$$w_{ij} = \left(\sum_{i<j} \delta_{ij}^2 \right)^{-1} \tag{6}$$

$$w_{ij} = \left(\sum_{i<j} \delta_{ij}^4 \right)^{-1} \tag{7}$$

With such weights the criteria (1) and (2) express relative error with respect to original dissimilarities enabling to compare the visualization precision of different data.

6. Testing Result

Different MDS algorithms for the same multidimensional data produce different two-dimensional images. Let us consider vertices of the unit hypercube in six dimensions whose dissimilarity is measured by Euclidean distance. The two dimensional image of the set of vertices is presented in Fig. 1.

Some regularity of the original data can be guessed from all images. The image produced by the *hybrid stress* algorithm shows the existence of different symmetries in the data as well as its substructures formed of 2^i points. The image obtained by the algorithm of the package STATISTICA is like a blurred version of the previously discussed image. Possibly the local optimization algorithm used by STATISTICA terminates prematurely. The other images presented in Fig. 1 have different shortcomings, e.g. in the image produced by hybrid MDS *sstress* each point shadows 15 other points.

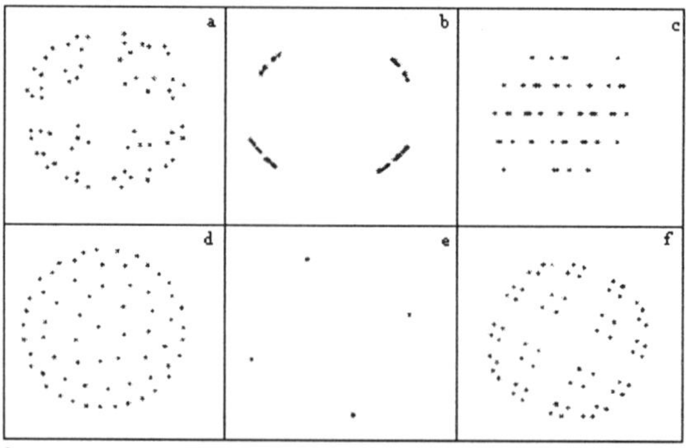

Figure 1. Images of vertices of six dimensional hypercube (dissimilarity is measured by *Euclidean* distance) produced by: a) STATISTICA, b) SPSS, c) NCSS, d) NCSS non-metric, e) hybrid, MDS *sstress*, f) hybrid, MDS *stress*.

The dissimilarity between the pairs of points measured by Minkowski distance is ordered in the same way as measured by Euclidean distance. However, the relative similarity can be different as shown by Fig. 2. Visualization results of the hypercube data with dissimilarities calculated according to Minkowski distance with $p = 1.5$, are presented at Fig. 3, are similar to those for the data with dissimilarities calculated according to Euclidean distance. However, the images obtained by means of non-metric NCSS algorithm in this case are more informative than in the case of Euclidean distances. Similar conclusion on Mahalanobis distance is justified by the images presented in Fig. 4. In the image obtained by means of hybrid MDS *sstress*, the points do not shadow the other points as in the Fig. 1(e), which makes Mahalanobis distance more reasonable to use. On the other hand, the dissimilarities calculated using the correlation coefficient (5) seem not appropriate for visualization (see Fig. 5). Only the image obtained by means of hybrid MDS *stress* seems more or less acceptable. From the images obtained by using correlation coefficient, we have chosen only those presentations, which tend to preserve typical regularity of the original data.

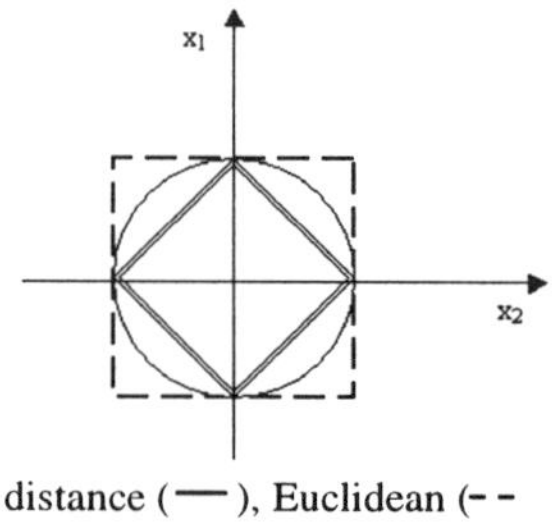

Figure 2. Unit circle in maximum distance (—), Euclidean (- -), and City block metrics ( ).

The structure of vertices of the hypercube is well exposed in the images obtained by the methods hybrid MDS *stress*, STATISTICA and NCSS. The dissimilarity measured by Mahalanobis distance seems most appropriate, because the corresponding data is well visualized by most algorithms. The two-dimensional data obtained by various methods explore the known structure. However, the best images are obtained by hybrid MDS *stress* method using Mahalanobis as well as Euclidean distances and by NCSS non metric method using Minkowski distance.

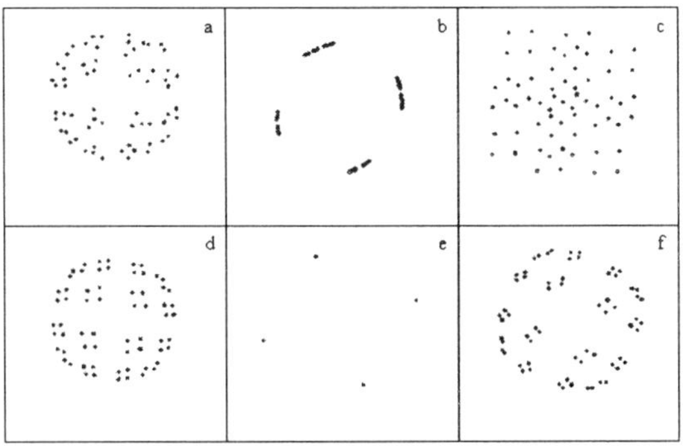

Figure 3. Images of vertices of six dimensional hypercube (dissimilarity is measured by *Minkowski* distance) produced by: a) STATISTICA, b) SPSS, c) NCSS, d) NCSS non-metric, e) hybrid MDS, sstress, f) hybrid MDS, stress.

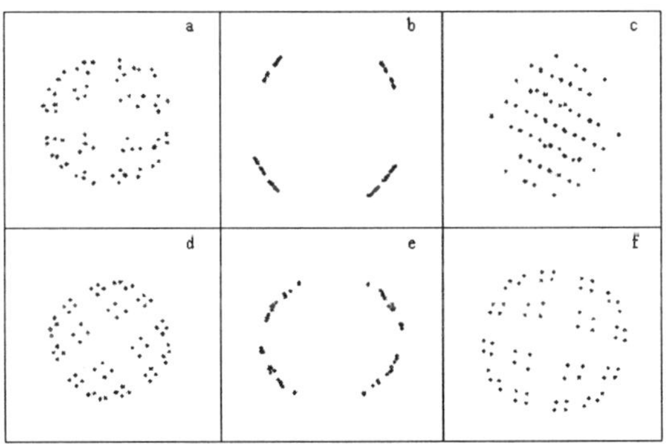

Figure 4. Images of vertices of six dimensional hypercube (dissimilarity is measured by *Mahalanobis* distance) produced by: a) STATISTICA, b) SPSS, c) NCSS, d) NCSS non-metric, e) hybrid MDS, *sstress*, f) hybrid MDS, *stress*.

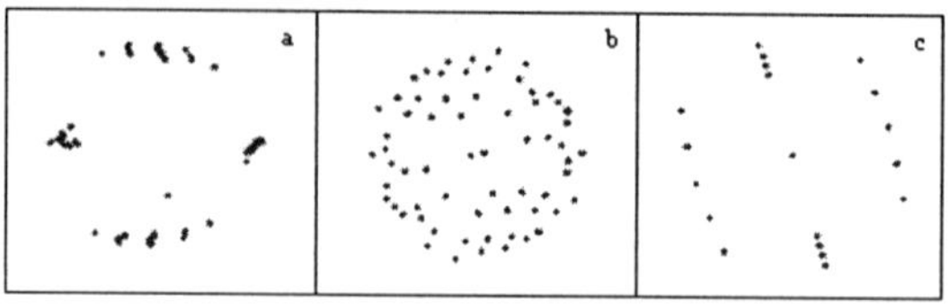

Figure 5. Images of vertices of six dimensional hypercube (dissimilarity is measured using *correlation coefficient*) produced by: a) SPSS, b) NCSS non-metric, c) hybrid MDS, *sstress*.

Visualizing Iris data gives analogous results, which had been already obtained by prior researches of the data.[12] As can be seen in Figs. 6 and 7, one class of two-dimensional points can be separated linearly from the other two. The two remaining classes, however, are somewhat entangled. In order to make full overview of the data structure in two-dimensional embedding space, we have analyzed the results produced by all above described visualization methods (5). STATISTICA, SPSS and hybrid MDS *sstress* methods presented most appropriate data structures composing three more or less distinguishable clusters. Detected clusters correspond to three analyzed Iris species that are have been chosen to analyze. In the image obtained by random search MDS method using Mahalanobis distance (see Fig. 7(b)) have achieved the biggest inter clusters distances. The other tested MDS methods give similar mapping results tending to expose two well separable classes.

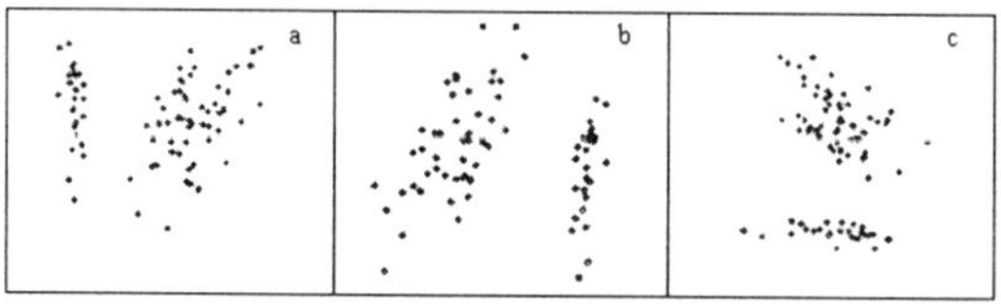

Figure 6. Images of Iris data (dissimilarity is measured by *Euclidean* distance) produced by: a)STATISTICA, b) SPSS, c) hybrid MDS, *stress*.

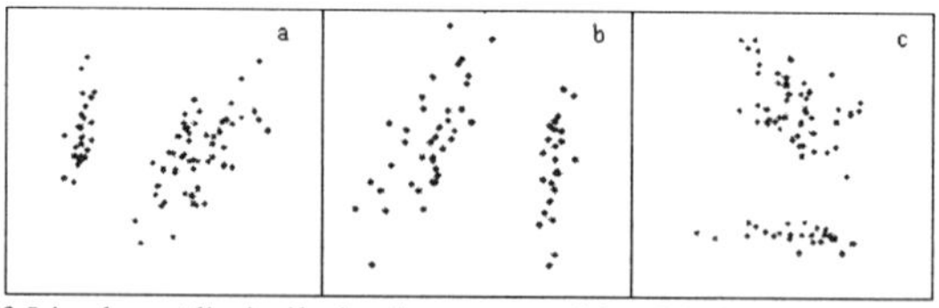

Figure 7. Images of Iris data (dissimilarity is measured by *Minkowski* distance) produced by: a) STATISTICA, b) SPSS, c) hybrid MDS, *stress*.

Visualization of biomedical data on sleep quality using different algorithms and different dissimilarities measures can be also summarized in favor of

Mahalanobis distance. The images obtained by SPSS, NCSS and hybrid MDS *stress* with Euclidean and Mahalanobis similarity measures are presented in Figs. 8, 9, 10 and 11. The mapping results distinguished three clusters corresponding to three subsets of patients: with good sleep quality, suffering from some sleep disorders, and with serious illness.

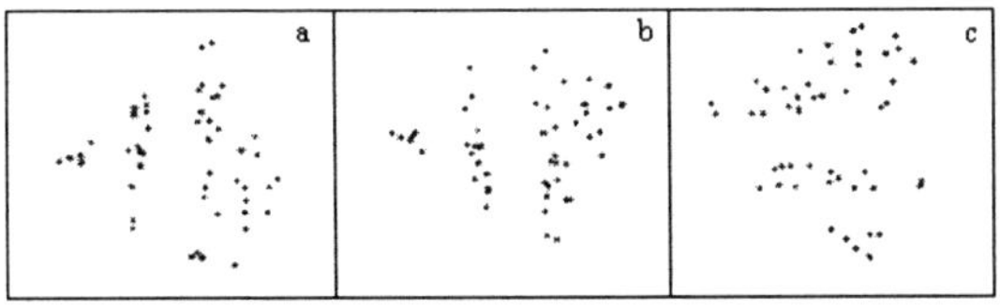

Figure 8. Images of data sets taken from "subjective sleep assessment" (dissimilarity is measured by *Euclidean* distance) produced by: a) SPSS, b) NCSS, c) hybrid MDS, *stress*.

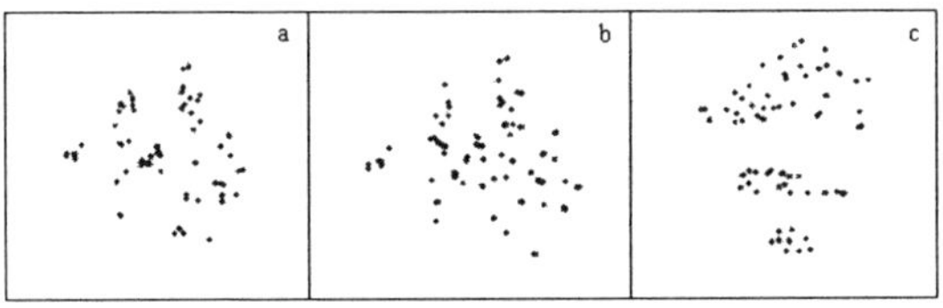

Figure 9. Images of data sets taken from "subjective sleep assessment" (dissimilarity is measured by *Minkowski* distance) produced by: a) SPSS, b) NCSS, c) hybrid MDS, *stress*.

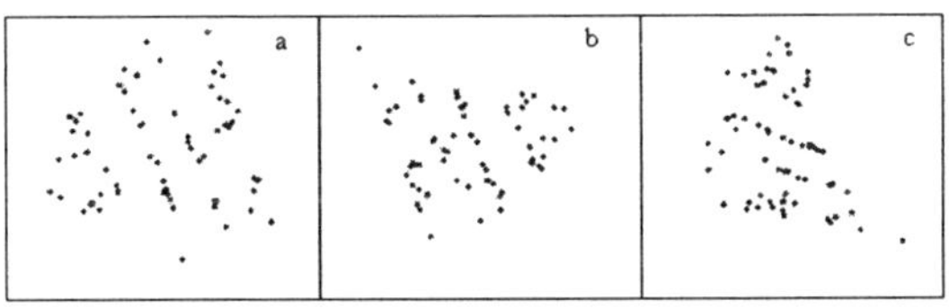

Figure 10. Images of data sets taken from "objective sleep assessment" (dissimilarity is measured by *Euclidean* distance) produced by: a) SPSS, b) NCSS, c) hybrid MDS, *stress*.

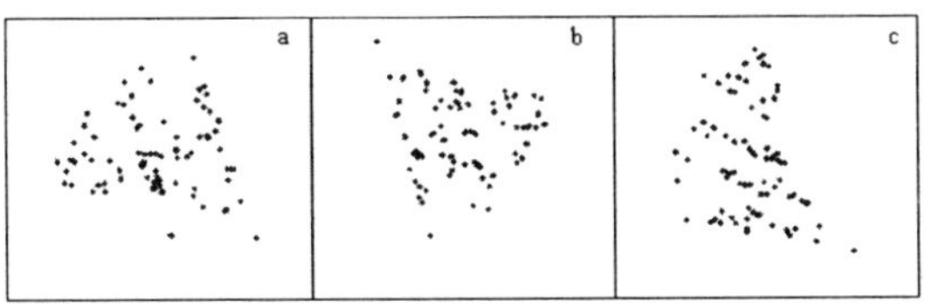

Figure 11. Images of data sets taken from "objective sleep assessment" (dissimilarity is measured by *Minkowski* distance) produced by: a) SPSS, b) NCSS, c) hybrid MDS, *stress*.

7. Conclusions

1) The metrics of the original space used as dissimilarity measure by MDS methods remarkably influences the informativeness of the images.
2) The structure of the original data sets is best exposed by two-dimensional images obtained by MDS methods using Euclidean and Mahalanobis distances.

Acknowledgement

The first author acknowledges the support of Lithuanian State Science and Studies Foundation.

References

1. U. Fayyad *et al.* (eds.), *Information Visualization in Data Mining and Knowledge Discovery*, Morgan Kautman, London/San Francisko (2002).
2. C. Jones, *Visualization and Optimization*, Kluwer (1995).
3. H. Hotelling, *J. Educ. Psychol.* **24**, 417-441, 498-520 (1933).
4. T. Kohonen, *Self-Organizing Maps*, Springer (1997).
5. I. Borg and P. Groenen, *Modern Multidimensional Scaling*, Springer (1997).
6. T. F. Cox and M. A. A. Cox, *Multidimensional Scaling*, Chapman and Hall/CRC (2001).
7. Ch. Bishop, M. Svensen and Ch. Williams, *Neural Computation* **10**(1), 215-234 (1998).
8. A. Torn and A. Zilinskas, Global optimization, *Lecture Notes in Computer Science* **355**, 1-255 (1989)
9. R. Mathar and A. Žilinskas, *Acta Applicandae Mathematicae*. **33**, 109-118 (1993).
10. P. Groenen, *The Majorization Approach to Multidimensional Scaling*, DSWO Press, Leiden (1993).
11. A. Zilinskas and A. Podlipskytė, *Information Technology and Control*, **3**(24), 49-54 (2002).
12. R. A. Fisher, *Annals of Eugenics 7*, 179-188 (1936).
13. A. Rechtschaffen and A. Kales, *A Manual of Standardized Terminology, Techniques, and Scoring for Sleep Stages in Human Subjects*, Wash. D.C.: U.S. Government Printing Office. 64 (1968).
14. D. J. Buysse *et. al.*, *Psychiatry Research*. **28**, 193-213 (1988).
15. A. Podlipskyte *et al.*, *Proceedings of Sixth International Conference Pattern Recognition and Information Processing, Minsk.* 181-188 (2001).

CORRECTION OF DISTANCES IN THE VISUALIZATION OF MULTIDIMENSIONAL DATA

J. BERNATAVIČIENĖ

Institute of Mathematics and Informatics, Akademijos g. 4
LT-08663 Vilnius, Lithuania

V. ŠALTENIS

Vilnius Pedagogical University, Studentų St. 39
LT-08106 Vilnius, Lithuania

A method for correcting interpoint distances before projecting on a lower dimensional space is presented in this paper. The basic idea is to change the distances according to the distribution of distances in high dimensional areas. Such corrections increase the quality of mapping: they distinguish the clusters of data points. The data structure after projection is more precise in this case. The proposed corrections are simple enough; the values of the correction coefficients were calculated for various data dimensionalities.

1. Introduction

Objects from the real world are frequently described by an array of parameters, *i.e.*, we deal with multidimensional data. A set of objects is usually presented in the form of a table or matrix. Each row of the table represents a different object. In fact, each row is a multidimensional vector (point, data item). One of the ways of analyzing such data is visualization.

In this paper, we discuss the visualization of multidimensional data by using structure preserving projection methods. These methods are based on the idea that the multidimensional data points can be projected on a lower dimensional space so that the structural properties of the data are preserved as faithfully as possible. Examples of such techniques are principal component analysis,[1] multidimensional scaling (MDS),[2] Sammon's mapping,[3] and others. When visualizing the multidimensional data by the nonlinear projection method, it is endeavored to distort the distances of data least. Each structure preserving the projection method has its own distortion criteria (loss or stress function). For MDS the general least squares loss function with transformed distances[4] is:

$$E_{MDS} = \sum_{i=1}^{s}\sum_{j=1}^{s} w_{ij}(f((d_{ij}^{*})^2) - f((d_{ij})^2))^2 . \quad (1)$$

In our case d_{ij}^{*} is the distance between two *n*-dimensional points $X_i, X_j \in R^n$, $i, j = 1,...,s$, instead of dissimilarity measure δ_{ij}; d_{ij} is the distance between two points $Y_i, Y_j \in R^2$ in a two-dimensional space, corresponding to the points $X_i, X_j \in R^n$, w_{ij} is fixed nonnegative weight. The function f(z) transforms the distances. Three particular examples of f(z) are presented:[4] $f(z) = z^{\frac{1}{2}}$ gives Kruskal's Raw Stress function[5] $f(z) = z$ gives S-Stress,[6] and $f(z) = \log(z)$ gives Ramsay's Multiscale loss function.[7]

Even having optimized distortion criteria exactly, the point projections after this type of nonlinear mapping often make the visual analysis difficult. One of the reasons for that are the different distributions of interpoint distances in high-dimensional areas.

The distributions of distances between two random points uniformly distributed in cubes were analyzed.[8,9] The comparison of such distribution for dimensionalities n=1,2,3 was presented[9] and the theoretical distributions were derived.[10] The transformation of distances in a three-dimensional cube was proposed in such a way that the distribution in three-dimensional cube would be the same as that in a two-dimensional cube.

In this paper, we propose another transformation function that is based on statistically estimated distribution. The transformation is suitable for higher dimensionalities. Such corrections may increase the quality of mapping: they distinguish the clusters of data points; the data structure after projection may be more precise and detect outliers.

In our investigations, the results of visualization with the correction and without it have been compared. Two distortion criteria classical multidimensional scaling[11] and Sammon[3] were used.

Classical multidimensional scaling (MDS) is the simplest nonlinear projection method, where the stress function is the sum of squared deviations of the distances:

$$E = \sum_{\substack{i,j=1 \\ i<j}}^{s} (d_{ij}^{*} - d_{ij})^2 . \quad (2)$$

Sammon's mapping is closely related to MDS. It also tries to optimize the stress function that characterizes how exactly the pair wise distances in a data set are preserved. The stress function of Sammon's mapping is the following distortion of projection:

$$E_s = \frac{1}{\sum\limits_{\substack{i,j=1 \\ i<j}}^{s} d_{ij}^*} \sum_{\substack{i,j=1 \\ i<j}}^{s} \frac{(d_{ij}^* - d_{ij})^2}{d_{ij}^*}, \tag{3}$$

here $\sum\limits_{\substack{i,j=1 \\ i<j}}^{s} d_{ij}^*$ is a constant normalizing factor.

2. Influence of Distance Correction on the Visualization Results

Every nonlinear distance correction has some influence on the projection quality.

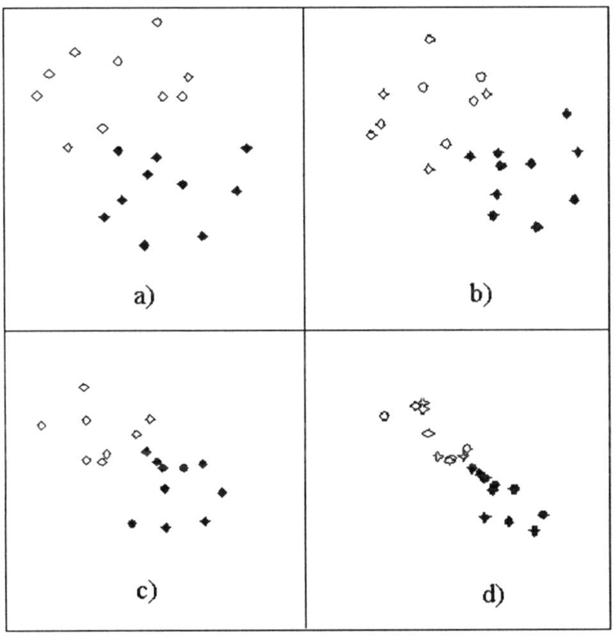

Figure 1. Projection of 6-dimentional uniformly distributed data in the two partly (13%) overlapped spheres with various correction degrees: a) without correction; b) correction $d_n^{1,2}$; c) correction $d_n^{1,6}$; d) correction d_n^3.

Figure 1 presents the projection of 6-dimensional uniformly distributed data in two partly (13%) overlapped spheres using various degrees of correction (projection has been obtained using MDS): a) without correction; b) with correction $d_n^{1,2}$; c) with correction $d_n^{1,6}$; d) with correction d_n^3; here d_n are distances in an *n*-dimensional space. After this type of correction the near points are more compressed.

The best result is obtained using the correction $d_n^{1,6}$ (see Fig. 1c), however the correction d_n^3 (see Fig. 1d) is too strong, because the large distances between the points are expanded too much.

The type of distance correction and its degree changes the quality of projection essentially. Therefore it is advisable to find the improving correction of distances. When determining requirements for the improving correction, we take into account the distribution particularities of the distances between points in high-dimensional areas.

3. Distribution of Distances Between Uniformly Distributed Points in a Unit Multidimensional Hypercube

Distributions of Euclidean distances between uniformly distributed points in a unit hypercube are investigated.[12] The probability density functions for dimensions n have been obtained by numerical simulation. Figure 2 illustrates these results.

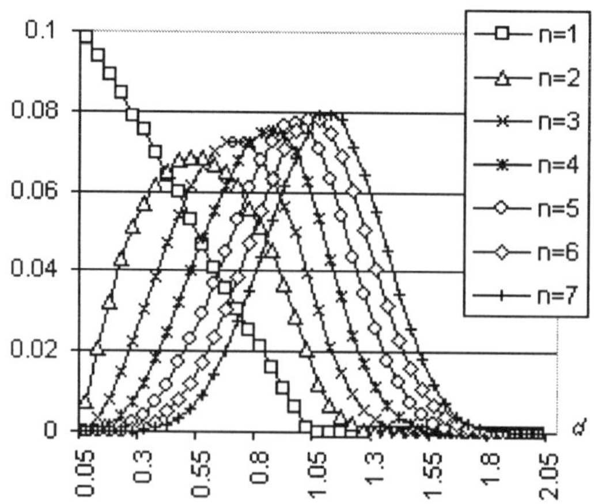

Figure 2. The probability density function for distances between the uniformly distributed multidimensional points in a unit hypercube for various dimensions.

We apply a distribution function $F_n(x)$, corresponding to the probability density of distribution $F_n(x)=P(\xi<x)$.

Here $P(\xi<x)$ is dependence of the probability on the given value x, when the distance ξ between random points is less than the given value x.

4. The Basic Idea of the Improving Correction

The monotony requirement is necessary for the improving correction. The distant data points must remain far after correction as well.

The correction must be such that the distances d_n between the uniformly distributed points in an n-dimensional hypercube would have the same distribution of distances d_2 as in a two-dimensional space (Fig. 3). To this end, the distances d_n are shortened multiplying them by the corresponding correction coefficient $k_n = \frac{d_2}{d_n}$. Here $F_n(d_n) = F_2(d_2)$, $F_n(d_n)$, $F_2(d_2)$ - distribution functions for *n*- and two-dimensional spaces.

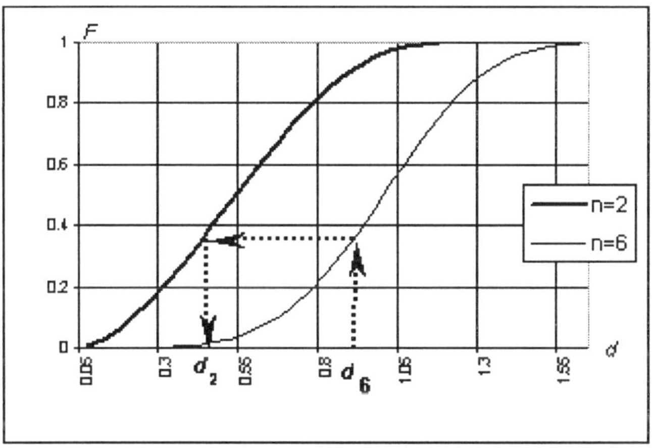

Figure 3. Example of the correction coefficient evaluation. Here the distribution functions of distances for two- and six-dimensional spaces are presented; d_6 is the value of distance for 6-dimensional space, d_2 is the corresponding distance after correction for a 2-dimensional space.

Thus we obtain the correction coefficient values for various distances d_n in an n-dimensional space (the distances are normalized in the interval [0,1]).

The function (4) approximates these dependences well enough.

$$k_d = 1 - \exp(-c_1(d_n - c_2)); \tag{4}$$

here the values of coefficients c_1 and c_2 are presented in Table 1.

Table 1. Values of coefficients c_1 and c_2.

Dimension n	Value of the coefficient c_1	Value of the coefficient c_2
4	1.4	0.04
5	1.18	0.16
6	1.05	0.25

When computing the coefficient k_d (4), it is useful to apply the values of coefficients c_1 and c_2.

Table 2. Values of the correction coefficient k_d or various dimensions n.

d_n	n			
	3	4	5	6
0	0	0	0	0
0.2	0.475	0.225	0.125	0.075
0.4	0.663	0.413	0.250	0.150
0.6	0.775	0.550	0.383	0.275
0.8	0.844	0.663	0.525	0.406
1	0.880	0.760	0.635	0.540
1.2	0.900	0.796	0.717	0.646

The approximation functions (4) are more convenient in comparison with the exact correction coefficient from the tables, but the approximation functions are not suitable for small values of distances (d_n<0.2). In this case, the coefficient values are used from the tables (Table 2). Figure 4 illustrates the transformation function of six-dimensional distances to two-dimensional ones. The dotted parts of the function were obtained by extrapolating and the values are approximate.

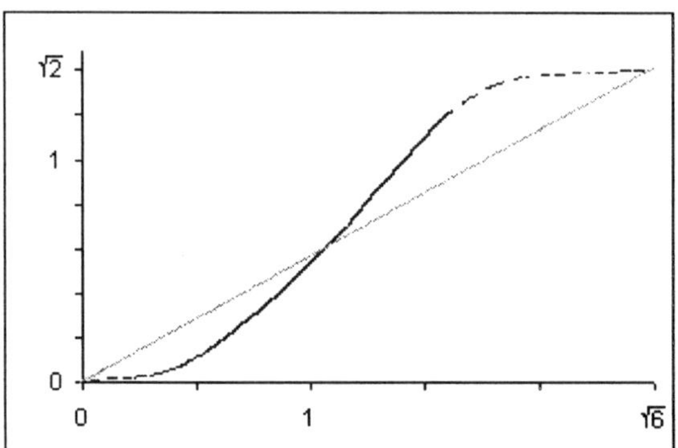

Figure 4. Transformation function of six-dimensional distances to two-dimensional distances.

5. Experimental Estimation of the Correction Results

Several local optimization methods were used to optimize the distortion criteria. The aim was not to compare them, but to investigate the quality of visualization results with correction and without the improving correction.

5.1. *Experiments with MDS*

A popular Quasi-Newton method[13,14] is used for the local minimization of the function, which characterizes the quality of visualization. The local optimization process is repeated with different initial points 30 times for global minimization.

Data used for the experiments

Two spherical data clusters of the same radius are not overlapping or partly overlapping. Each cluster consists of 10 data points.

In Fig. 5, we can see the illustration of the mapping results where the spherical data clusters overlap (by 40%). The certain data clusters without correction are not separable (see Fig. 5a), but after the correction we can visually separate two clusters (Fig. 5b).

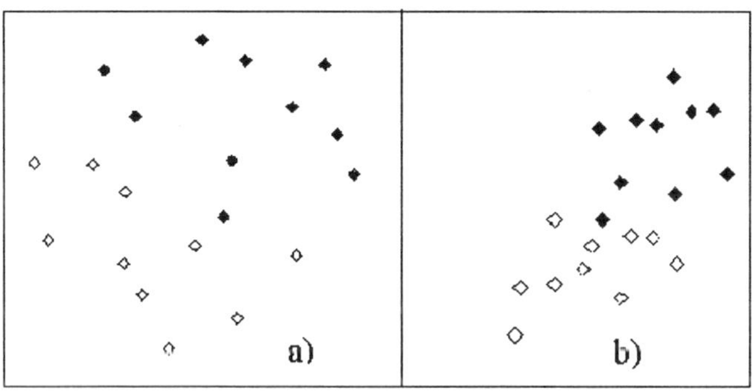

Figure 5. Projections of 6-dimensional uniformly distributed points in two overlapped spheres: a) without correction; b) after the improving correction.

Projections of two spherical data clusters, which are remote from each other, are illustrated in Fig. 6. The clusters are precisely separated in both cases: without correction and after the improving correction. However the projection of the clusters without correction of distances is scattered. The data points seem more clustered after the improving correction.

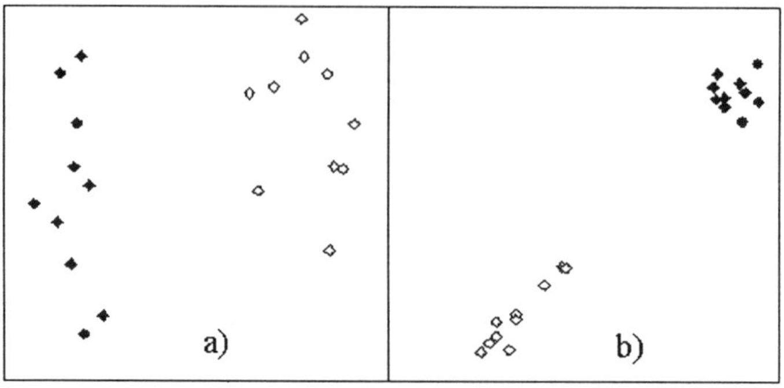

Figure 6. Projections of 6-dimensional uniformly distributed points in two not overlapping spheres: a) without correction; b) after the improving correction.

5.2. *Experiments with the Modified Sammon Method*

The Sammon algorithm, in which Seidel-type coordinate descent[15] is applied, is used in the following experiments. The local optimization process is repeated with different initial points 100 times for global minimization.

Data used for the experiments

- Two spherical not overlapping data clusters of the same radius. Each cluster consists of 20 data points.
- The Wood data set.[16] The data set consists of 20 five-dimensional points. Four points are outliers and form a separate cluster.

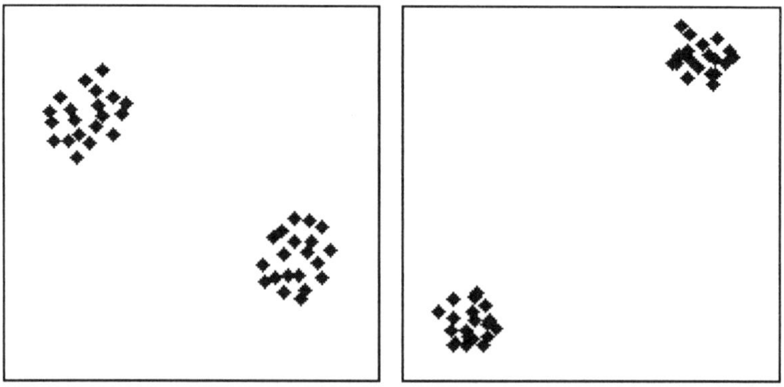

Figure 7. Projections of 6-dimensional uniformly distributed points in two not overlapping spheres: a) without correction; b) after the improving correction.

The projections of two spherical data clusters, which are remote from each other, are illustrated in Fig. 7: a) without correction; b) after the improving correction. The correction does not influence the projection result considerably. However, the figures show that, after the improving correction, the structure of clusters becomes clearer.

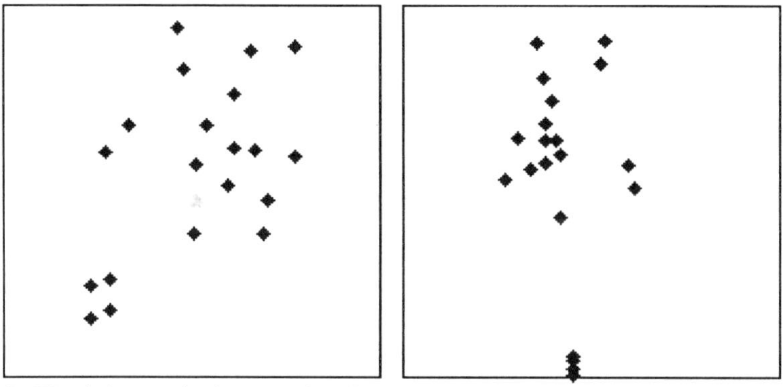

Figure 8. Wood data projections on the plane: a) without correction; b) after the improving correction.

The projections of Wood data without correction are presented in Fig. 8a. It is difficult to separate four outliers in this case. However, after the improving correction, near points are compressed (Fig. 8b) and the four points - outliers look like one cluster.

6. Conclusions

The method of correction of interpoint distances is presented. The basic idea is to change the distances according to the distribution of distances in high dimensional areas. The values of the correction coefficients were calculated for various data dimensionalities. In order to demonstrate the quality of visualization, the experiments have been done on some datasets. Multidimensional scaling and modified Sammon mapping methods were used.

The proposed corrections are simple enough. It is necessary to multiply the pair wise distances in a multidimensional space by the respective corresponding correction coefficients, which are different for various dimensionalities. The table of the correction coefficient values and approximated functions are presented for several dimensionalities.

In the future research, a comparative analysis of several transformation of this kind will be investigated and their influence on visualization results will be explored more in detail.

References

1. P. Taylor, Statistical Methods, In: M. Berthold and D. J. Hand (eds.), *Intelligent Data Analysis: an Introduction*, Springer-Verlag, 69-129 (2003).
2. I. Borg and P. Groenen, *Modern Multidimensional Scaling: Theory and Applications*, Springer, New York (1997).
3. J. W. Sammon, A nonlinear mapping for data structure analysis, *IEEE Transactions on Computers* **C-18**, 401-409 (1969).
4. P. J. F. Groenen, R. Mathar and J. De Leeuw, Least squares multidimensional scaling with transformed distances, In: W. Gaul and D. Pfeifer (eds.), *Studies in Classification, Data Analysis, and Knowledge Organization*, Springer, Berlin, 177-185 (1996).
5. J. B. Kruskal, Multidimensional scaling by optimizing goodness of fit to a nonmetric hypothesis, *Psychometrika* **29**, 1-27 (1964).
6. Y. Takane, F. W. Young and J. de Leeuw, Nonmetric individual differences in multidimensional scaling: An alternating least squares method with optimal scaling features, *Psychometrika* **42**, 7-67 (1977).
7. J. O. Ramsay, Maximum likelihood estimation in MDS, *Psychometrika* **42**, 241-266 (1977).
8. Sh. Sabirov, Distribution function of the distance between two points inside a cube, *Random Oper. and Stoch. Equ.*, **8**, 339-342 (2000).
9. A. Zilinskas and A. Podlipskyte, On symbolic computation in problem of geometric probabilities, *Information Technology and Control* **24**, 49-54 (2002).
10. A. Zilinskas, On the distribution of the distances between two points in a cube, *Random Operators and Stochastic Equations* **11**, 21-24 (2003).
11. W. S. Torgerson, Multidimensional Scaling, I: theory and method, *Psychometrika* **17**, 401-419 (1952).
12. V. Saltenis, Outlier detection based on the distribution of distances between data points, *Informatica* **15**(3), 399-410 (2004).
13. W. C. Davidon, *Variable Metric Method for Minimization*, A.E.C. Research and Development Report, ANL-5990 (1959).
14. R. Fletcher and M. J. D. Powell, A rapidly convergent descent method for minimization, *Computer Journal* **6**, 163-168 (1963).
15. G. Dzemyda, J. Bernatavicienė, O. Kurasova and V.Marcinkevicius, Minimization of the mapping error using coordinate descent, In: *Conference proceedings WSCG'2005,* January 31-February 4, 2005, Plzen, ISBN 80-903100-9-5, Czech Republic, 169-172 (2005).
16. N. R. Draper and H. Smith, *Applied Regression Analysis*, John Wiley and Sons, New York (1966).

FORECASTING OF BANKRUPTCY WITH THE SELF-ORGANIZING MAPS ON THE BASIS OF ALTMAN'S Z-SCORE

E. MERKEVIČIUS

Informatics Department, Kaunas faculty of Humanities, Vilnius University, Muitines str.8, Kaunas, Lithuania

In financial institutions statistical and artificial intelligence methods have been used for determination of the credit risk classes. Recently, algorithms of the artificial neural networks were often applied, one of them is a *self-organizing map* (SOM): this is the two-dimensional map of the credit units in the process that is generated by similar characteristics (attributes), however, this process is not specified by network outputs. If in the cluster dominate credit units of one class, and then it is reasonable to use SOMs in forecast of bankruptcy of the company. Here was used the core statistical methodology on prediction for bankruptcy of corporate companies created by Altman so called Z-score model. The factors of bankruptcy were used as input data and Z-score values were used to define clusters of generated SOM. The results of our investigations were presented and they show that SOM is reliable method for bankruptcy prediction.

1. Introduction

The forecasting of the credit state has always been a relevant task in the finance market. Available algorithms of statistical and artificial intelligence methods especially the combinations of them adduce more and more accurate predictable results. Recently has been employed the algorithms of artificial neural networks. Artificial neural networks are divided into supervised and unsupervised learning.[1]

Self-organizing map (SOM) is unsupervised learning artificial neural network that is generated without defining of output values. As a result, in two-dimensional clusters map are visualized credit units by similar characteristics. The purpose of this paper is to investigate the capabilities of SOM in forecasting of credit classes.[2]

Martin-del-Prio and Serrano-Cinca have been one of the first who applied SOM to the financial analysis. They generated SOMs of the Spanish banks and subdivided them into two large groups and this allowed establishing root causes of the banking crisis.[1]

Kiviluoto[3] made a map by means of including 1137 companies, 304 companies from them were crashed. SOMs are said to give useful qualitative information for establishing similar input vectors. Visual exploration allows seeing the distribution of important indicator – bankrupt – on the map, thus, it is possibly to apply that map for the forecasting of-companies bankrupt.

The following authors have estimated oniy a current situation of credit state and afterwards they have interpreted it for forecasting bankrupt, causes of crisis period or market segmentation of banks. In this paper, we suggest generating SOM as one that could be applied for forecasting of credit classes for new companies.

The first section describes Z-score bankruptcy model by Altman and the core steps of standard SOM algorithm. In the second part we present the results of our investigations with real credit data and propose several recommendations for generating valuable SOM employing to credit class forecast.

2. Methodology

In the self-organizing process output data are configured to visualization of topologic original data.[4,5,6] The learning of SOM is based on competitive learning ("winner takes all"). The algorithm of standard stochastic SOM learning is described.[2] This algorithm is realized in various software with refined capabilities.[4] In our investigations we chose SOM software in that is realized above SOM algorithm.

Altman's Z-score predicts whether or not a company is likely to enter into bankruptcy within one or two years. Original Altman's model has been presented.[7,8] We used original Z-score model for private firms.[8] A healthy private company has a $Z > 3$; it is non-bankrupt if $2.7 < Z < 2.99$; it is in the watch-listed zone if $1.8 < Z < 2.69$; it is unhealthy (bankrupt) if it has a $Z < 1.79$.

3. Investigations

The aim of these investigations is to explore and estimate possibilities of SOM method predicting bankruptcy level of Lithuanian companies. The basis for generating of SOM is financial dataset from EDGAR PRO Online database. Calculated bankruptcy ratios are used as inputs and Z-scores by Altman's Z-Score model are used as labels for identification of characterized areas (clusters) in the SOM.

3.1. *Description of Datasets*

In this paper possibilities of SOM's is studied by means of two real financial datasets: companies from NASDAQ list, (or TRAINDATA) and Lithuanian companies (TESTDATA). We chose Viscovery SOMINE ®[9] which is characterized as a suitable and simple interface, large possibilities of data pre- and post processing, adopted Batch SOM, fast learning rate, comprehensive visualization and monitoring tools.[10]

Table 1. Characteristics of financial datasets.

Dataset	TRAINDATA	TESTDATA
Taken from	EDGAR PRO Online Database (free trial)	one of Lithuanian commercial bank database.
Period of financial data	2004Y	2004Y
Count of records	1110	742
Number of inputs (attributes)	5	
Risk class of bankruptcy	0–3 (healthy–bankrupt), according to Altman Z-Score model for private companies	

3.2. *Evaluation*

The algorithm of experiment contains 4 steps:

1.Getting financial ratios (inputs) from financial datasets. There are used 5 ratios of bankruptcy for private companies by Altman's Z-score method for calculation of bankruptcy coefficient Z (or Z-score). Z-scores are converted to 4 risk classes: if Z is less than 1,8 then is assigned 3 risk class (bankrupt), Z-score in the interval [1,8–2,7] means 2 risk class (watch-listed), Z-score in the interval [2,7–3] means 1 risk class (non-bankrupt) and if Z is at least 3 then risk class is 0 (healthy). In the Table 2 the statistical description of inputs of TRAINDATA and TESTDATA datasets is showed:

Table 2. Statistical description of inputs.

Inputs (weights in Z-score model)	Datasets	Min	Max	Mean	Std. deviation
EBIT/Total Assets (3,107)	TRAINDATA	3,167	2,783	-0,013	0,620
	TESTDATA	-1,52	2,491	0,224	0,322
Net Sales /Total Assets (0,998)	TRAINDATA	0	6,717	0,793	0,849
	TESTDATA	0	6,797	1,732	1,099
Market Value of Equity / Total Liabilities (0,42)	TRAINDATA	-0,249	8,367	0,929	1,214
	TESTDATA	-0,136	6,897	0,495	0,707
Working Capital/Total Assets (0,717)	TRAINDATA	-0,664	0,702	0,071	0,386
	TESTDATA	-0,721	0,665	0,105	0,166
Retained Earnings /Total Assets (0,847)	TRAINDATA	-6,933	1,046	-0,413	1,134
	TESTDATA	-0,406	0,798	0,349	0,184

We see that all inputs in the datasets have similar statistical characteristics, except Retained Earnings/Total assets. The minimal value of that ratio in TRAINDATA equal to -6,933 whereas in TESTDATA equal to -0,406.

2. Generating SOM from TRAINDATA.

3. The estimating of network training reliability shows how values of output (risk class) distribute in clusters of SOM. Appropriate number of clusters is estimated according to distribution of risk classes in the areas of SOM. In the Fig. 1 the SOM of TRAINDATA is presented:

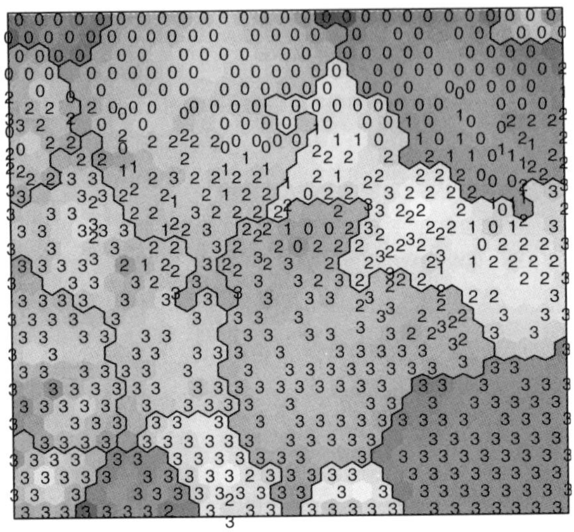

Figure 1. Generated SOM with labels of risk classes.

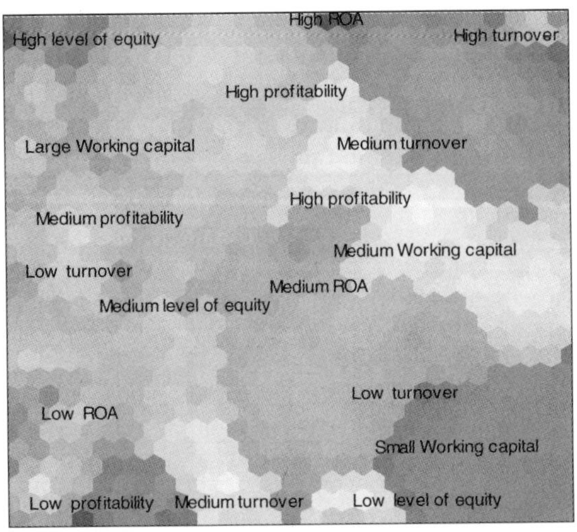

Figure 2. SOM with characteristic of areas.

It can be seen that each cluster includes only one dominant risk class.

4. Testing of trained SOM. On the trained SOM the credit class values of TESTDATA are putted on. It is estimated the distribution of credit class values on the trained SOM clusters and is compared with means of credit classes values of TRAINDATA. In Fig. 3 the distribution of TESTDATA in SOM is presented:

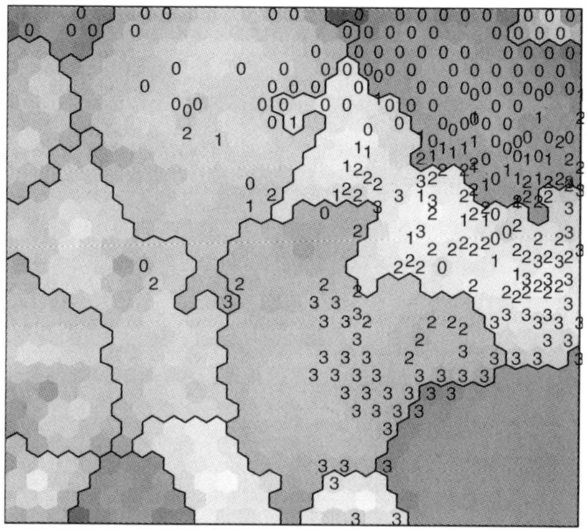

Figure 3. Distribution of Lithuanian companies risk classes on the SOM.

In the Table 3 the means of distribution of credit classes in the SOM is presented:

Table 3. Distribution of risk classes of Lithuanian companies.

	RISK CLASS				*Dominant*	*Distribution in*	*Grand*
Cluster	*0*	*1*	*2*	*3*	*risk class*	*cluster (in %)*	*Total*
C 1	31	2	3	1	0	83,78	37
C 12	8				0	100,00	8
C 13				4	3	100,00	4
C 14	38				0	100,00	38
C 2	3		18	74	3	77,89	95
C 3	22	17	113	70	2	50,90	222
C 4	214	41	68	6	0	65,05	329
C 5	1		1		2	50,00	2
C 6				7	3	100,00	7
Grand Total	317	60	203	162		81%	742

We see that the vast majority of records distributed in the 3 clusters: "C4", "C3" and "C2".

Half of total records in the Cluster "C4" are distributed. This cluster is characterized as high turnover, high profitability, medium amount of working capital, low level of equity. Cluster "C4" is labeled as "healthy cluster" because app. two-third (65,05%) records in this cluster have Z-score values above 3, i.a. healthy companies.

In the Cluster "C3" one-third records of total dataset is distributed. This cluster characterized as "watch-listed" (credit class assigned to 2), but 31,5% of total records in this cluster are assigned to 3 risk class. So, this cluster have not unambiguous characteristic. Cluster "C2" have measurable large amount of dominant records of 3 risk class (77.89%) and this cluster could be labeled as cluster of bankrupt companies.

The average of distribution in clusters seeks 72.76%. If clusters labels of TESTDATA will correspond to clusters labels of TRAINDATA, then we can say that probability of prediction of risk classes seeks exactly 72.76%.

Comparing of TRAINDATA and TESTDATA distribution in the clusters of SOM is showed in the Table 4.

Table 4. Comparing of TRAINDATA and TESTDATA distribution in the clusters of SOM.

Clusters	Weight of cluster (in %)	TRAINDATA		TESTDATA	
		Dominant risk class	Distribution in cluster (in %)	Dominant risk class	Distribution in cluster (in %)
C 1	14,05	0	70,51	0	83,78
C 10	1,98	3	95,45	N/A	
C 11	2,43	3	100,00	N/A	
C 12	2,16	0	100,00	0	100,00
C 13	1,17	3	100,00	3	100,00
C 14	1,08	0	100,00	0	100,00
C 2	13,24	3	76,87	3	77,89
C 3	11,53	2	53,13	2	50,90
C 4	11,80	0	69,47	0	65,05
C 5	7,03	3	87,18	2	50,00
C 6	23,15	3	100,00	3	100,00
C 7	4,77	3	100,00	N/A	
C 8	2,88	3	93,75	N/A	
C 9	2,70	2	43,33	N/A	
Grand Total			85%		84,70%

It can be seen that in the cluster "C5" dominant risk class of TRAINDATA do not correspond to dominant risk class of TESTDATA, so we must eliminate this cluster from general calculation of prediction probability.

Eliminating Cluster "C5" lets narrowly improve probability of prediction to 84,70%.

4. Concluding Remarks

The capabilities of SOM in the prediction of bankruptcy of corporate companies were investigated in this article. Several techniques and recommendations were proposed which could give more reliable results in using SOM in this area. In the investigation the probability of Lithuanian corporate companies' prediction of bankruptcy reached 84,70%.

For prediction of more reliable results in the future we will change several attitudes to investigation:

- via calculation methodology of Altman`s Z-score: we expect, that by changing weights of factors prediction probability will increase and this model will be adopted to the conjuncture of Lithuanian companies;
- exploration of financial data will be period of two years;
- by expanding amount of TRAINDATA and TESTDATA records, the quality of neural network training and testing will be improved.

References

1. B. Martin-del-Prio and K. Serrano-Cinca, Self-organizing neural network: the financial state of Spanish companies, In: *Neural Networks in Finance and Investing*, 341-357 (1993).
2. E. Merkevičius, R. Simutis and G. Garšva, Forecasting of credit classes with the self-organizing maps, *Information Technology and Control* **4**(33), 61-66 (2004).
3. K. Kiviluoto, Predicting bankruptcies with the self-organizing map. *Neurocomputing* **21**, 191–201 (1998).
4. G. Deboeck, Financial applications of self-organizing maps. *American Heuristics Electronic Newsletter*, Jan (1998).
5. T. Kohonen, The self-organizing map, *Proceedings of the IEEE* **78**, 1464-1480 (1990).
6. T. Kohonen, Self-organized formation of topologically correct feature maps, *Biological Cybernetics* **43**, 59-69 (1982).
7. E. Altman, Financial ratios, discrimination analysis and the prediction of corporate bankruptcy, *Journal of Finance* **23**, (1968).

8. E. Altman, Predicting financial distress of companies: revisiting the Z-score and ZETA® models, http://pages.stern.nyu.edu/~ealtman/Zscores.pdf (2000).
9. Viscovery SOMine. Eudaptics software Gmbh, http://www.eudaptics.at.
10. G. Deboeck and T. Kohonen, *Visual Explorations in Finance with Self-Organizing Maps*. Springer Finance, London (1998).

THE MOST APPROPRIATE MODEL TO ESTIMATE LITHUANIAN BUSINESS CYCLE*

A. JAKAITIENĖ
Mathematical Statistics Department, Vilnius Gediminas Technical University, Sauletekio 11, 10223 Vilnius, Lithuania
Macroeconomics and Forecasting Division, Bank of Lithuania, Gedimino 6 01103 Vilnius, Lithuania, E-mail: ajakaitiene@lb.lt

This paper reviews four methods of estimating the output gap in Lithuania, including Hodrick-Prescott (HP) filter, Prior-Consistent (PC) filter, production function model and a multivariate unobserved components models. Latent variables obtained using Kalman filter. All estimates of output gap show that the economy of Lithuania has been above it's potential level at the end of 2004. The Kalman filter output gap was less volatile, compare to other methods and only from smooth results of Kalman filter it was possible to identify Lithuanian economy business cycle. A long-run potential growth of the Lithuanian economy is estimated at 5.75 per cent. We could not prove that the Kalman filter reduced end-of-sample uncertainty. Kalman filter was always underestimating the output gap, but HP filter tended to overestimate.

1. Introduction

Modelling trends and cycles in time series has a long history in empirical economics. Economic fluctuations have presented a recurring problem for economists and policymakers starting the end of the nineteenth century till latter day. Rather than cycles, they tended to think in terms of "crisis", used to mean either a financial panic or period of deep depression.[1] However, today we understand business cycle as economy-wide fluctuations in output, incomes and employment.[2] A revision of the recent research on the estimation of the business cycles reveals a wide range of methodologies. However, it is possible to identify two basic approaches: structural and non-structural methods. Structural methods are based on

*The author wish to thank I. Vetlov and R. Kuodis for their helpful comments. The views expressed in this paper are those of the author and do not necessarily represent those of Bank of Lithuania.

economic theory, and they include methodologies such as the production function approach,[3,4,5] structural vector autoregressions,[6,7,8] Okun Law's method[9] and, recently, general equilibrium models.[10] On the contrary, non-structural methods are purely statistical and mechanic in its nature without any explicit economic foundation. Methodologies under this second approach go from the very simple linear and segment approaches to more complex ones as the Baxter and King or Band Pass filters,[11] and the very popular Hodrick-Prescott (HP) filter. There is also a mixed approach that combines properties of structural and non-structural methods. Known as multivariate methods, they introduce economic relationships to statistical de-trending methods. This class includes the multivariate Hodrick-Prescott Filter,[12,13] the Beveridge-Nelson Decomposition, the Unobserved Components method[14,15,16] and multivariate structural models using Kalman filter.[17] Essentially, all methods rely on the assumption that each observed variable at a certain period can be decomposed into cyclical component (business cycle or gap) and trend component (potential level).

Lithuanian business cycle was already estimated in research related to nominal and real convergence of new European member states to Eurozone (ECB[18], EFN[19], MNB[20]). The authors applied variety of filtering procedures for two observed variables: the quarterly GDP and monthly industrial production. Also there are a few research papers in those the production function approach was used to estimate potential output growth.[3,4] The latter reference is the Lithuanian Block of the European System of Central Banks (ESCB) Multi-Country Model (MCM). In the model aggregate supply is approximated by Cobb-Douglas production function. The Bank of Lithuania uses this model for generating forecasts. The estimates of potential output growth, as results of HP filter and production function, is used by Lithuanian Ministry of Finance as well.[21] In the research carried out by both institution, it is agreed that univariate filters do not fully reflect some structural changes of economy and suffer from the end-of-sample problems. As a result we would be interested in multivariate filter or in identifying business cycle as common cycle of such variables as output, unemployment, inflation. A multivariate model would allow to use information contained in the price series to assist the decomposition of the unemployment rate. Consequently, this specification would be potentially closer to the non-accelerating inflation rate of unemployment concept than univariate techniques[9]. This could provide us with better understanding of economy and would guarantee some degree of consistency with the model.

This paper reviews four methods of estimating business cycle or output gap for Lithuanian economy, including HP filter, PC filter, production function model and a multivariate unobserved components models. In the latter model, potential output, the non-accelerating inflation rate of unemployment, the effect of monetary policy and several other parameters are determined simultaneously, which enables their interactions to be taken into account. Also we would expect that multivariate unobserved component model, as suggest the empirical research, will be more stable and it will reduce end-of-sample uncertainty compare to HP filter. The applied structural model behind the unobserved components method was selected similar to the experience of Czech National bank.[17] Unobserved variables obtained using Kalman filter. Production function was chosen as Cobb-Douglas production function with constant returns to scale and labour-augmenting technological process (as it is used in[4]). Since real GDP is the best measure of economic well-being we used it for approximation of output.

The rest of the paper is organized as follows. Section 2 presents the econometric techniques considered to estimate potential output and output gap. Following that, in Sec. 3, we will represent our empirical part. The paper will end with results and conclusions.

2. Theoretical Framework

The business cycle is typically defined as deviation from trend (or potential level). Many kinds of filtering have been proposed in order to decompose output into its high and low frequency components which were respectively assigned to output gap and potential output. The detrending methods considered in this paper are: The Hodrick-Prescott filter, The Prior-Consistent filter, Unobserved Component model, Production Function.

Next we briefly introduce each of four methods.

2.1. *The Hodrick-Prescott (HP) Filter*

The univariate Hodrick-Prescott filter would be a classical way to estimate the business cycle and it could be called as a "first guess". The method is a simple smoothing procedure that has become increasingly popular because its flexibility in tracking the characteristics of fluctuations in trend output.

Formally, the potential output $\overline{x}_t^{HP}$ using the HP filter is derived from

$$\arg\min \sum_{t=1}^{T} (x_t - \overline{x}_t^{HP}) + \lambda^{HP} \sum_{t=2}^{T-1} \left[(\overline{x}_{t+1}^{HP} - \overline{x}_t^{HP}) - (\overline{x}_t^{HP} - \overline{x}_{t-1}^{HP}) \right]^2 .$$

However, HP filter could be presented as state-space model for any variable x as well. The measurement equation for HP filter would be as follows $x_t = \bar{x}_t^{HP} + xgap_t^{HP}$ and the transition equations equal

$$\begin{aligned} \bar{x}_t^{HP} &= 2\bar{x}_{t-1}^{HP} - \bar{x}_{t-2}^{HP} + \varepsilon_t^{\bar{x}^{HP}} \\ xgap_t^{HP} &= \varepsilon_t^{xgap^{HP}}, \end{aligned}$$

where $\bar{x}_t^{HP}$ stands for the potential level of x, $xgap_t^{HP}$ is the x gap, $\varepsilon_t^{\bar{x}^{HP}}$ and $\varepsilon_t^{xgap^{HP}}$ are disturbances.

The covariance matrix of the error terms is

$$Q = \begin{pmatrix} \sigma^2_{\varepsilon^{\bar{x}HP}} & 0 \\ 0 & \sigma^2_{\varepsilon^{xgapHP}} \end{pmatrix} \text{ or } Q = \begin{pmatrix} \frac{\sigma^2_{\varepsilon^{\bar{x}HP}}}{\sigma^2_{\varepsilon^{xgapHP}}} & 0 \\ 0 & 1 \end{pmatrix} = \begin{pmatrix} \frac{1}{\lambda^{HP}} & 0 \\ 0 & 1 \end{pmatrix} .$$

λ is called the "smoothing parameter" and penalizes the variability in the growth component. The larger the value of λ,the smoother the growth component and the greater the variability of the output gap. As λ approaches infinity, the growth component correspond to a linear time trend. For quarterly data, Hodrick and Prescott propose setting λ equal to 1600.

2.2. *The Prior-Consistent (PC) Filter*

This univariate filtering methodology is similar to the HP filter, with the advantage that it allows for the imposition of priors on the properties of series (such as the levels, rates of change or variance).[22] Estimates of potential output $\overline{x}_t^{PC}$ from the PC filter are derived as

$$\arg\min \sum_{t=1}^{T} (x_t - \overline{x}_t^{PC}) + \lambda^{PC} \sum_{t=2}^{T-1} \left[(\overline{x}_t^{PC} - \overline{x}_{t-1}^{PC}) - \Delta\overline{x}_t^{PC*} \right]^2 .$$

$\Delta\overline{x}_t^{PC*}$ denote some prior estimate of the change in the variable.

In the language of state-space models, the PC filter would be described in very similar way as HP filter above. The initial values of state variables were set to the value of the first observation. The initial gap was set to zero. The parameter λ^{PC} has been fixed to 25 in all applications (see Box 7 in[22]). We assumed that if a "large" deviation for the trend were equal to 1, then the corresponding variance value in gap terms would be 5.

2.3. *Unobserved Component Model*

The applied structural model behind the unobserved components method was selected similar to the experience of Czech National bank.[17] In analysis the following structural model was applied

$$y_t = \bar{y}_t + ygap_t \tag{1}$$

$$u_t = \bar{u}_t - ugap_t \tag{2}$$

$$p_t = (1 - \alpha_1 - \alpha_2) \cdot p_{t-1} + \alpha_3 \cdot ygap_{t-1} + \tag{3}$$

$$+ \alpha_1 \left[\Delta_4 p_t^M + 100 \cdot \Delta_4 r_t^{eq}\right] + \alpha_2 \pi_t + \varepsilon_t^{\pi} \tag{4}$$

$$\bar{y}_t = \bar{y}_{t-1} + \mu_{t-1} - \omega_1 \cdot \Delta \bar{u}_t + \varepsilon_t^{\bar{y}_t} \tag{5}$$

$$\mu_t = \gamma_1 \cdot \mu_{t-1} + (1 - \gamma_1)\, \bar{\mu} + \varepsilon_t^{\mu} \tag{6}$$

$$ygap_t = \theta_1 \cdot ygap_{t-1} - \theta_2 \cdot igap_t - \theta_3 \cdot rgap_t + \varepsilon_t^{ygap} \tag{7}$$

$$\bar{u}_t = \bar{u}_{t-1} + \varepsilon_t^{\bar{u}} \tag{8}$$

$$ugap_t = \beta_1 \cdot ugap_{t-1} + \beta_2 \cdot ygap_t + \varepsilon_t^{ugap} \tag{9}$$

Equation (1) is identity that simply defines that output y_t is a composition of the log of potential level of output $\bar{y}_t$and the output gap $ygap_t$. Equation (2) is another identity which states that the unemployment rate u_t is equal to difference between NAIRU $\bar{u}_t$and the unemployment gap $ugap_t$. According this definition we would expect positive correlation with the output gap measure. In (3) the inflation respond to expected inflation π_t and annual changes in import prices p_t^M corrected for proxy of the equilibrium level of log real exchange rate r_t^{eq}. Equation (5) determines the dynamics of potential output $\bar{y}$. The equation of the growth rate of potential output μ_t is described as a first order, stationary autoregressive process. Equation (7) describes the dynamics of the output gap. The gap is formulated to evolve according to a first-order autoregressive process, but allowing for effects from the real average loans interest rates gap $igap_t$ and the real exchange rate gap $rgap_t$. Real average loans interest rates were calculated as difference between nominal interest rates and annual inflation. Equation (8) specifies the NAIRU as a pure random walk driven by shocks $\varepsilon_t^{\bar{u}}$. Last equation is an Okun equation that links the movements in the unemployment gap those in the output gap. Δ_4 means the four quarter difference operator.

The ε_t^{π}, $\varepsilon_t^{\bar{y}_t}$, ε_t^{μ}, ε_t^{ygap}, $\varepsilon_t^{\bar{u}}$, ε_t^{ugap} variables are random variables that are assumed to be identically, independently and normally distributed and to be uncorrelated.

2.4. *Production Function*

In contrast to the structural model, this approach represents a supply-side view of the economy. In particular, the structural model is appropriate for detecting the current inflationary pressure, but is of less use in efforts to discuss either the factors of potential output growth (technological progress, capital and labour) or future potential output developments stemming from the evolution of the supply side. Production function is chosen as Cobb-Douglas production function with constant returns to scale and labour-augmenting technological process. This type of function approximates aggregate supply in the Lithuanian Block of the ESCB MCM.[3] The Bank of Lithuania uses this model for generating forecasts. Function itself equals to

$$y_t = k_t^{\beta} l_t^{(1-\beta)} a_0 \exp\left[(1-\beta)\gamma T\right]$$

where k_t - real capital stock, l_t - employment, β- constant that measures capital's share of income, γ- rate of technological progress, T – linear trend.

It is assumed that actual capital stock is close to potential and potential level of employment equals to employment that corresponds to the natural rate of unemployment or NAIRU. With that potential level of production and output gap are calculated as follows

$$\bar{y}_t^{PF} = k_t^{\beta} \bar{l}_t^{(1-\beta)} a_0 \exp\left[(1-\beta)\gamma T\right], \; ygap_t^{PF} = \frac{y_t - \bar{y}_t^{PF}}{\bar{y}_t^{PF}}$$

where $\bar{l}_t$- potential employment consistent with natural rate of unemployment or NAIRU.

For modelling quarterly data is used. The data sample is from 1st quarter of 1997 to 4th quarter of 2004.

3. Empirical Analysis

As the structural model was already described, we have to estimate or calibrate it. There is a permanent dilemma which way to choose as the "truth" model does not exist. The Lithuanian data sample is very short and covers a period of structural changes in the economy and changes in policy regimes. Therefore we had chosen the calibration. To begin the calibration, we need some measures of key exogenous "gaps" (as r_t^{eq}, $igap_t$, $rgap_t$). For this the univariate PC filter was used.

In our analysis the calibrated parameters were used similar to Czech National bank.[17] Their calibrated values presented in the first row of Table 3, as the estimated values are in the second row. The estimated parameters for the economy of Lithuania are in the third row. Almost all estimated parameters are close to calibrated values with two exceptions (α_2 and α_3). From the theoretical point of view, the presumably positive output gap should also feed into inflationary pressures, but our past experience shows that in very small open economies this link tends to be rather weak, as most of the extra domestic demand tends to channel into extra quantities imported. α_3 was set to 0.15 instead of 0.5 as we have not found any significant correlation between proxy of output gap (obtained using PC filter) and annual inflation (excluding energy) and α_2 was set to 0.1.

Table 1. Calibrated parameters.

Parameter	$\widehat{\alpha}_1$	$\widehat{\alpha}_2$	$\widehat{\alpha}_3$	$\widehat{\theta}_1$	$\widehat{\theta}_2$	$\widehat{\theta}_3$	$\widehat{\beta}_1$	$\widehat{\beta}_2$
Value	0.2484	0.33	0.5	0.90	0.25	0.15	0.9	0.1
Estimation_CZ	0.2676	0.106	0.43	0.92	0.19	0.05	0.88	0.08
Estimation_LT	0.14	0.30	0.05	0.62	0.10	0.14	0.64	0.17

Parameter ϖ_1 was set at 0.4, based on the approximate share of labour income in total income. In economy with large structural changes we would not expect very quick convergence to it's steady-state level and the parameter γ_1 was set at 0.9. The technical description of the measurement and transition equations for the structural model described above is available.[23] The structural model is estimated by maximum likelihood using the support of the Kalman filter.

On behalf of estimating production function, we face a problem of capital stock time series for Lithuania. Therefore initial capital stock was estimated using regression analysis within the Perpetual Inventory Method.[3] However the potential level of employment consistent with natural rate of unemployment or NAIRU was obtained applying PC filter.[3] From production function approach we calculated that estimate of parameter β is equal to 0.55 for Lithuania. This indicates that capital share of income is 55 per cent as labour creates another 45 per cents.

The calculations was performed in Eviews 4.1.

4. Results

We attempted to estimate Lithuanian economy business cycle using four different methods in 1997-2004. All estimates of output gap show that the economy of Lithuania has been above it's potential level at the end of 2004 (see Table 4). The duration of being above it's potential level differ in results of Kalman and all other methods. Following HP and PC filters and production function approach the Lithuanian economy has been above it's potential already from the first quarter of 2003. But Kalman filter gives somewhat different picture. From the 3rd quarter of 2002 absolute value of output gap becomes positive and fluctuates around zero for the next four quarters. The significant change in the 3rd quarter of 2003 coincides with the vigorous annual growth of GDP (close to 11 per cent). Also the Kalman filter output gap is less volatile, compare to other methods (Table 4). In essence, only from Kalman filter smooth results it is possible identify Lithuanian economy business cycle as other output gap estimates are too fluctuating. From this we identified that Lithuanian economy is in it's third business cycle. Full business cycle took just over 4 years (17 quarters) in the analysis period. As there is no data available before 1997 it is impossible to discuss the cycle which ended up by 4th quarter of 1998 though the economy started to decline from the later and was falling out till the third quarter of 2000. This was consequence of Russian crisis when industry and agriculture export has significantly decreased to Russia.[24] After this recession the economy started to recover and it took only 5 quarters to reach positive output gap.

Table 2. Estimated annual growth of potential output, in per cent.

	Average			Standard deviation		
	98-00	00-04	98-04	98-00	00-04	98-04
HP filter	4.61	6.73	5.91	0.22	0.78	1.22
PC filter	4.12	6.82	5.78	0.60	0.76	1.51
Kalman filter	5.01	6.22	5.76	2.83	2.80	2.82
Production function	5.30	6.47	6.02	1.33	1.54	1.55
Memo: real GDP	2.37	7.62	5.60	4.04	2.12	3.92

As we already presented the Kalman filter, as most stable method, compare to other methods, the end-of-sample problem is still open. Some of literature and research agree[17,1] that all de-trending methods suffer of end-of-sample problem, though estimates obtained using Kalman filter has smaller end-of-sample uncertainty compare to the other filters. We have checked

Table 3. Annual output gap, in per cent of potential level.

Method	1997	1998	1999	2000	2001	2002	2003	2004
Kalman filter	3.0	1.6	-2.2	-4.8	-3.9	0.0	0.2	0.7
HP filter	1.3	4.0	-1.9	-2.1	-1.5	-1.4	1.6	1.1
PC filter	0.8	3.0	-2.1	-1.8	-1.3	-1.5	1.2	1.1
Production function	2.4	4.8	-3.2	-2.9	-1.1	-1.2	1.3	1.2

the stability of our results for Kalman and HP filters. For this reason we iterated our model three times with different samples: the end of 1st sample was 4th quarter of 2002, the second – the end of 2003 and the third – 2004. The picture looks puzzled. For the 1st sample we obtained that Kalman filter is more stable as difference at the end of 2002 was twice smaller then HP filter (accordingly 1.1 and 2.9 percentage points). However, the changes in the results of the 2nd sample were on the HP filter side. The fairly short time series could be the reason for this instability. Meanwhile, we should pay attention that Kalman filter always underestimate the output gap, but HP filter tends to overestimate.

5. Conclusions

Lithuanian business cycle was estimated using four different methods, including HP filter, PC filter, production function model and a multivariate unobserved components models. Unobserved variables were obtained using Kalman filter. All estimates of output gap show that the economy of Lithuania has been above it's potential level at the end of 2004. The duration of being above it's potential level differ in results of Kalman and all other methods. The Kalman filter output gap was less volatile, compare to other methods and only from smooth results of Kalman filter it was possible to identify Lithuanian economy business cycle. From this we identified that Lithuanian economy is in it's third business cycle. Full business cycle took just over 4 years (17 quarters) in the analysis period. Also we have found that the potential growth in two different periods (till 2000 and after) was different between selected methods. Univariate filters showed almost twice larger difference between growth in two periods. This means that inflation and monetary variables had dampened effect on the average growth of potential output. As a result, we might think about 5.75 per cent growth as long-run potential growth of the Lithuanian economy.

We have checked the stability of our results for Kalman and HP filters with different sub-samples. We could not prove that the Kalman filter reduced end-of-sample uncertainty. Kalman filter was always underestimating the output gap, but HP filter tended to overestimate.

Finally, despite the advantages of multivariate Kalman filter with the structural model behind, it faces disadvantages as well. Any error in the model structure or in the size of elasticities can be critical for output gap identification. In addition, the choice of noise variances is critical for the identification results. Based on our practical experience with the Kalman filter, we believe that the advantages of this multivariate filtering technique outweigh its potential drawback. Equally we would not recommend to trust on one method results, but all these methods use as complementary tools.

References

1. T. C. Mills, *Modelling Trends and Cycles in Economic Time Series*, Palgrave MacMillan (2003).
2. N. G. Mankiw, *Macroeconometrics*, Worth Publishers (2000).
3. I. Vetlov, Bank of Lithuania, *Monetary Studies* **No. 3**, 14–34 (2003).
4. I. Vetlov, *BOFIT Discussion Papers* **No. 13** (2004).
5. *CNB Economic Research Bulletin*, No.1, **Vol. 3** (2005).
6. R. Ramos and J. Suriñach, mimeo, University of Barcelona (2004).
7. J. Fidrmuc and I. Korhonen, *Comparative Economic Studies* **46**(1), 45-62 (2004).
8. I. Claus, **DP2000/03**, Reserve Bank of New Zealand (1999).
9. F. Orlandi and K. Pichelmann, *Economic Papers* **No. 140**, ECFIN (2000).
10. A. Trigari, *Working Paper Series* **No. 304**, ECB (2004).
11. J. Lawrence and T. J. Fitzgerald, *International Economic Review* **v44**, 435-465 (2003).
12. O. Basdevant, **DP2003/02**, Reserve Bank of New Zealand (2003).
13. D. Laxton and R. Tetlow, Technical Report **No. 59**, Bank of Canada (1992).
14. J. Hamilton, *Time Series Analysis*, Princeton University Press, Princeton (1994).
15. J. Durbin and S. J. Koopman, *Time Series Analysis by State Space Methods*, Oxford University Press (2001).
16. A. Scott, **DP2000/04**, Reserve Bank of New Zealand (2000).
17. Czech National Bank, *The Czech National Bank's Forecasting and Policy Analysis System* (2003).
18. ECB, *The New EU Member States Convergence and Stability* (2004).
19. M. Artis, M. Marcellino and T. Proietti, *EFN, Autumn 2003 Report* (2003).
20. Z. Darvas and G. Szapary, *MNB Working paper* **2004/1** (2004).
21. *Convergence Programme of Republic of Lithuania 2005-2008*, Ministry of Finance (2005).
22. D. Laxton, P. Isard, H. Faruqee and others, *IMF OP* **164** (1998).
23. A. Jakaitienė, *Lithuanian Mathematical Journal* **Vol. 45**, 486-493 (2005).
24. Bank of Lithuania, *Annual report* (1999).

EVALUATING THE APPLICABILITY OF TIME TEMPERATURE INTEGRATORS AS PROCESS EXPLORATION AND VALIDATION TOOLS

S. BAKALIS, P. W. COX, K. MEHAUDEN AND P. J. FRYER

Centre for Formulation Engineering, Dept. of Chemical Engineering, University of Birmingham, Birmingham, B15 2TT, UK

Knowledge of the impact of thermal processing in the food industry is crucial in order to deliver high quality safe foods to the consumer. Time Temperature Integrators (TTIs) have been developed as quality control and process exploration tools for processes where use of other thermal sensors is impossible. TTIs are encapsulated enzymatic suspensions with well characterized thermal inactivation kinetics, whose activity can be measured easily before and after processing. From the reduction of the TTI activity it is possible to estimate the inactivation of pathogens and spoilage organisms, as well as nutrients in the product. Although TTIs are currently used in many industries a thorough review of their applicability to evaluate thermal processes has not yet been published. Here, experimental validation of an α-amylase TTI is shown with the intention of accurately characterising the variability of the technique. In an attempt to describe the thermal variability of real food processes the heat and mass transport in typical food processes where TTIs might be used were simulated using CFD. Monte Carlo simulations to study the effect of (i) process variability and (ii) the measurement variability inherent within TTI response. Results indicate that TTIs can be used both to validate thermal processes; and as a process exploration tool. In the latter form, they can be used to derive information about variation, although a larger number of TTIs would be required.

1. Introduction

Food producers are responsible for the safety of the products that they are manufacturing. To guarantee the safety of their products, manufacturers use different preservation techniques. One most commonly used is to apply high temperatures to the food to reduce the quantity of heat sensitive micro organisms or spores present, which if not removed are responsible for food poisoning or food deterioration.[1]

At constant temperature, the micro organism population death is given by:[2]

$$\frac{N_{\text{final}}}{N_{\text{initial}}} = 10^{-t/D} \qquad (1)$$

$$D = D_{ref} 10^{\frac{T - T_{ref}}{z}} \tag{2}$$

Here N_{final} is the number of micro organisms after the heat treatment, $N_{initial}$ is the initial number of micro organisms, t is the heat treatment duration (minutes) and D is the decimal reduction time (in minutes) which is the heating time necessary to reduce an existing number of micro organisms by 90% at a specific temperature.[3] The effect of temperature on D value determination can be seen in (2), where D_{ref} is the decimal reduction point at the reference temperature (T_{ref}) and z is the temperature change that causes a log change in the decimal reduction time. The z is a function of the kinetics of the micro organism chosen as a target and it is typically obtained experimentally by measuring the kinetics of the microbial inactivation under thermal processing.

It is possible to compare different thermal treatments by evaluating an equivalent processing time at a reference temperature T_{ref}. Therefore, (1) becomes

$$F\,(\text{or } P) = D_{ref} \,.\log\left(\frac{N_{initial}}{N_{final}}\right) \tag{3}$$

$$P \text{ or } F = \int_0^t 10^{\frac{T(t) - Tref}{z}} .\,dt \tag{4}$$

Where F or P are the process values (F for sterilization and P for pasteurization). In order to calculate the process value, the organism under study should have inactivation behaviour which follows the linear thermal death time model described by.[4,5] If the temperature history of the product is known (4) can be used to estimate the F value.[2,6] It should be noted that (4) is a local linearization of more conventional Arrhenius kinetics; the equation works because food processing uses a relatively narrow temperature range.

Although a lot of effort has been devoted during the past decades towards obtaining a fundamental understanding and model thermal processing there seems to be an inability of the food industry to incorporate them in rationalizing their processes. Numerical simulations are often developed for idealised conditions ignoring the highly non linear rheological properties of food materials, therefore models are typically used to predict overall trends rather than optimizing possessing conditions. Furthermore, raw materials for the food industry are biological the variability in their physical properties is very high causing additional problems when deterministic modelling is use. Recently, the importance of this uncertainty has been acknowledged, for example Kronin *et al.*[7] investigated the effect of uncertainty of parameters such as sample dimension and heating rate on the thermal treatment. Further implications in the case of the food industry arise from microbial variability. The pathogen flora in food materials could vary significantly in terms of their thermal inactivation characteristics. Although significant efforts have been made towards developing models that account for this source of variability[8] the food industry commonly faces this problem by over processing in order to ensure safe products, with often detrimental results in terms of quality.

As previously mentioned, thermal processing is evaluated by experimentally determining the F value commonly by using thermocouples. However, thermocouples are not convenient for every type of thermal process. In some cases, they can interfere in the movement of the fluid and lead to incorrect time temperature history. In addition, their size can be an issue especially for access to some processing equipment since it has been impossible so far to miniaturize them plus they are mainly made from high heat conductible materials.[6,9]

Time Temperature Integrators (TTIs) are now being considered as an alternative to thermocouples. They are a relatively new technology that allows determining the impact of a process on a product attribute.[10] TTIs are devices having a height of approximately 2cm and a diameter of 0.5 cm which contain a thermally labile substance.

Under a heat treatment, the substance encapsulated inside the device undergoes irreversible change. These changes can be quantified before and after heat treatment and converted to a F or P value.[5,11,6,12,13,10] TTIs present many advantages over thermocouples; they are small, almost neutrally buoyant, can be made from materials which present the similar thermal characteristics as the target foods and the time temperature history of the product is not needed to determine the impact of thermal treatments.[5,6,10]

Although TTIs are valuable tools in the evaluation of process efficiency published literature on their performance and applicability is limited. In this work a both numerical and experimental tools were used to evaluate the applicability of TTIs and maximize the information obtained.

2. Materials and Methods

In this study, the enzyme used in the TTI preparation was the α-amylase (EC 3.2.1.1. Type II-A supplied by Sigma) isolated from *Bacillus amyloliquefaciens*. Amylase solution (20 μl of 10 mg/ml amylase is tris buffer) was encapsulated into pipes (Altesil high strength silicone tubing from Altec, 2 mm bore 0.5 mm wall) of 15 mm length. The extremities of each pipe were closed by silicone (Sylgard 170 elastomer). After heat treatment, the remaining amylase activity in the TTIs was measured by a standard spectrophotometric method.

2.1. *Estimation of TTI Parameters*

The D value was determined by measuring the remaining TTI activity after different lengths of times under isothermal conditions at different temperatures. The slope of these lines is the D-value at each temperature. The z value can be estimated by calculating D values at different temperatures and using (2). Figure 2 plots D versus temperature, and z is the slope of the line.

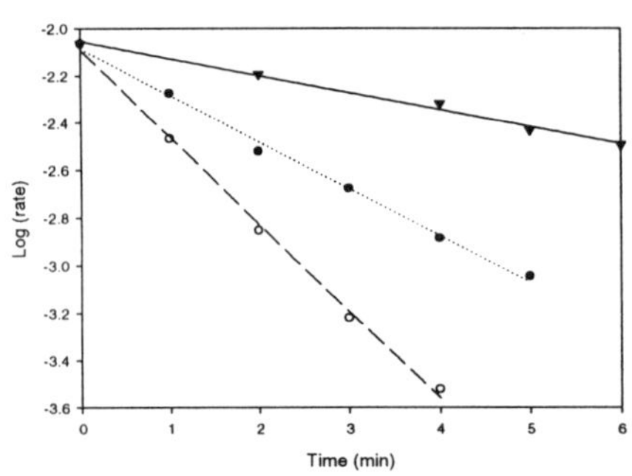

Figure 1. Inactivation curves used to estimate D value at — 80°C, ···· 85°C and – – 90°C.

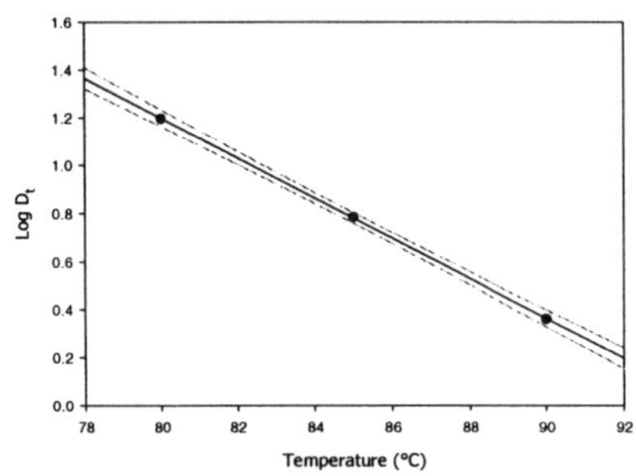

Figure 2. Plot of D vs. T used to estimate z.

Validation of the z and D_{ref} values was also performed at non-isothermal conditions.

2.2. *Estimation of the TTIs Variability Under Non Isothermal Heat Treatments*

The time temperature profiles used in this experiment were realised on a Peltier stage. The Peltier is a semiconductor which functions as a small heat pump; heat can be transferred from one side to the other of the thermoelectric module by applying a low voltage. Conveniently, this phenomenon is reversible and by changing the polarity of the system, the heat can be moved in the opposite direction to allow for cooling. The TTIs were given a heat treatment similar to those of a typical process. These experiments were repeated with a number of TTIs to define the reproducibility of the process.

2.3. *Numerical Simulation*

A critical question to address is that of how many TTI values are required to obtain sufficient understanding of the process to ensure safety of the product and gain understanding of the process. An idealized process relevant to that given by a tubular heat exchanger was modelled using CFD to develop the methodology and demonstrate principles of using TTIs. The model process consisted of a flow in a 2.5 m long pipe of radius 0.015 m. The velocity followed Hagen-Poiseuille flow (Newtonian viscosity independent of temperature was assumed). The temperature profile was assumed to be linear along the radial direction and constant along the axial direction. TTIs were assumed to follow flow stream lines and not to have any effect on the velocity distribution. The path lines were found by integrating velocity distributions. A Monte Carlo simulation was performed by introducing a variable number of TTIs in the process at various radii as shown in Fig. 3.

The radial distribution of TTIs was predicted by estimating the percentage of flow coming from each radial location. By consecutively following the path lines and the temperature history of TTIs, using (4) the P value for each TTI was predicted.

3. Results and Discussion, Estimation of the TTIs Variability

The histogram in Fig. 4 shows 54 P values obtained from different TTIs held for 4 min at 85°C.

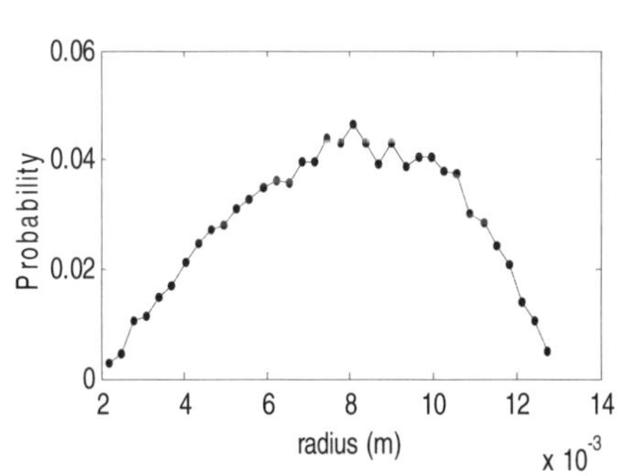

Figure 3. Radial distribution of TTIs.

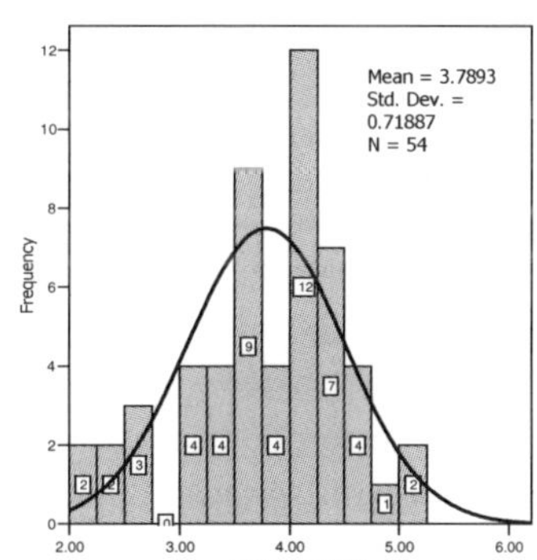

Figure 4. Variability of the TTIs responses for the same heat treatment (P=4).

The TTIs appeared to be normally distributed, with a mean P value of 3.79 min and a standard deviation of 0.72 min. The source of errors obtained with the

TTIs might be due to (i) variability in making the TTIs, (ii) variation of the position of the TTIs on the Peltier plate or (iii) the TTIs itself since it is of biological origin. Some extreme low P values appear on the histogram, they might be due to a poor contact between the TTIs and the Peltier plate. In Fig. 5, the variability of TTIs with increasing nominal P value is shown. TTIs appear to underestimate P.

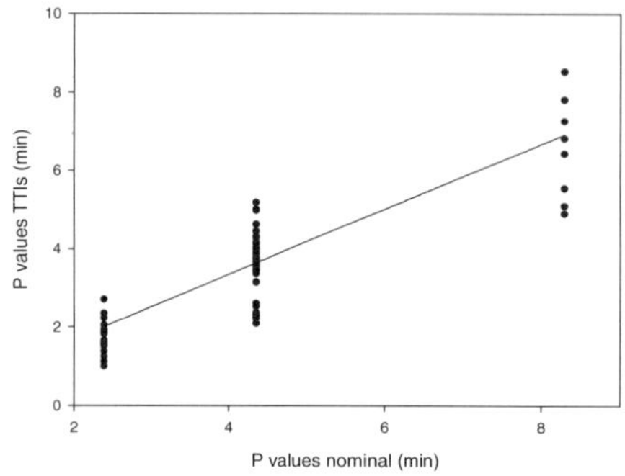

Figure 5. Responses of the TTIs at different processing levels.

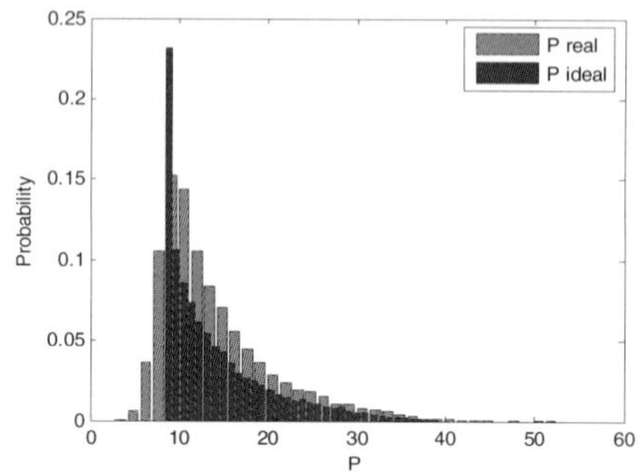

Figure 6. Distribution of P values in minutes for an ideal and a real system.

One can also see that the variability of the TTI prediction increases with increasing nominal P values. This behaviour was modelled by assuming that the TTIs followed a normal distribution with the standard deviation increasing linearly with nominal P values.

In Fig. 6, the distribution of P values from an ideal TTI system (where there was no variability between TTIs) are compared with P values from a realistic system (where there is some variability between TTIs).

For the system examined the uncertainty of the TTIs resulted in an increase in the spread of the resulting P values distribution. One of the main problems faced in evaluating a thermal process using TTI's, is determining the number of TTIs that should be used for any given piece of process plant. In Fig. 7, the resulting mean P value vs the number of TTIs used is shown. The solid lines correspond to two average values that can be estimated for the process. The lower line corresponds to an arithmetic mean; the top to a volumetric average obtained from the relationship

$$P_{vol-mean} = \frac{\int_0^r 2\pi r u(r) P(r) dr}{\int_0^r 2\pi r u(r) dr} \tag{5}$$

where u(r) is the velocity at a radial position r and P the P value of the TTI entering at that position. In principle the resulting P values from the TTIs are quite close to the mean and if more than 50 TTIs are used there should be sufficient confidence that a good estimation of the mean P value would be obtained. Furthermore, in real situations it is imperative to avoid under-processing of food materials, i.e. the minimum P value experienced from the material should dependent upon microbiological factors. In Fig. 8 the minimum P values of TTIs are plotted vs the number of TTIs.

Clearly the minimum value resulting from the TTIs is smaller than the value estimated from an ideal process due to the uncertainty in the TTI response which can result in values lower than the actual, thus under predicting the actual P values. Figure 8 shows that for the system under consideration, if more than 20 TTIs are used it is possible to accurately predict the minimum P values. Currently, in the food industry process designers tend to over process in order to ensure product safety (i.e. a minimum P value in the process) this results in an often unacceptable loss of quality. Development of the methods of application of TTIs and their incorporation into the design process will enable more rational thermal design to ensure safety without sacrificing quality.

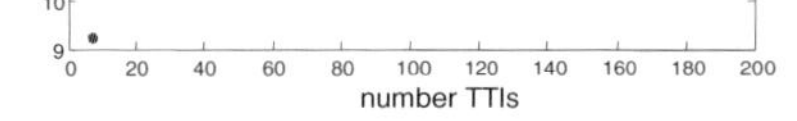

Figure 7. Mean and arithmetic mean P value for different number of TTIs.

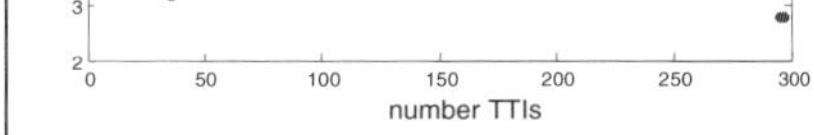

Figure 8. Minimum P values vs number of TTIs.

Significant interest exists in the food industry towards understanding and quantifying the variability in the heat treatment experienced by foods during processing. This might be quantifying by using the standard deviation of the P values, as measured by TTIs. In Fig. 9, the standard deviation against the number of measuring TTIs is shown.

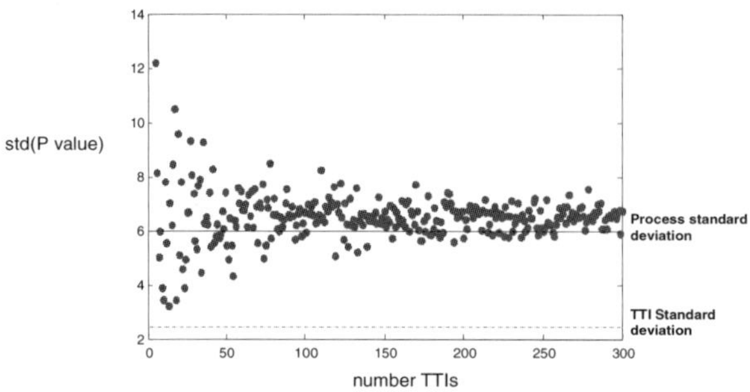

Figure 9. Standard deviation of P values vs number of TTIs.

The standard deviation value predicted from the TTIs is slightly higher than the one of the ideal process, perhaps due to the variability of the TTIs. It should be kept in mind that the number of TTI's required to make an accurate prediction of the standard deviation is much higher than the one need to evaluate thermal processing.

In the above example an idealized process with simplified temperature and momentum field was used. The same principle could be extended and evaluative programs developed that could be used with the currently available data forms and with results obtained from other sources such as commercial CFD applications or experimental results. Thus a framework that could be used during the design process for deciding upon the optimal heat treatment that a given process can deliver.

4. Conclusions

The food industry very commonly over processes products to ensure safety, but this might result in an unacceptable loss of quality. Novel tools such as TTIs could be used to rationalise process design and significantly contribute to the process optimisation. In this work a framework in which the number of TTIs used as well as the quality of the information that could be obtained was developed. It was found that approximately 50 TTIs could result in useful information about the mean and minimum thermal treatment experienced applied to the food; but a significantly higher number of TTIs may be necessary if information about the variability of thermal treatment is required.

Acknowledgements

Financial support from EPSRC and Giusti Ltd (for KM) and for DEFRA Link (for PWC) is gratefully acknowledged.

References

1. P. J. Fryer, D. L. Pyle and C. D. Rielly, *Chemical Engineering for the Food Industry*. Blackie Academic & Professional, London (1997).
2. C. Ball and F. Olson, *Sterilization in food technology*, New York (1957).
3. K. J. Valentas, E. Rotstein and R. P. Singh, *Handbook of Food Engineering Practice*. CRC Press LLC, Boca Raton (1997).
4. W. D. Bigelow, The logarithmic nature of thermal death time curves, *Journal of Infectious Diseases* **29,** 528-536 (1921).
5. Y. P. Guiavarc'h, F. T. Zuber, A. M. Van Loey, and M. E. Hendrickx, Combined use of two single-component enzymatic Time Temperature Integrators: Application to industrial continuous rotary processing of canned Ravioli, *Journal of Food Protection* **68,** 375-383 (2005).
6. M. E. Hendrickx, G. Maesmans, S. De Cordt, J. Noronha, A. M. Van Loey and P. Tobback, Evaluation of the integrated time temperature effect in thermal processing of foods, *Critical Reviews in Food Science and Nutrition* **35,** 231-262 (1995).
7. K. Cronin, K. Abodayeh, J. Caro-Corrales, A. Pokrovskii and A. Demir, Probabilistic studies of the thermalprocessing of discrete solid products, *Trans IChemE* **78**, 126-130 (2005).
8. S. T. Chou, L. T. Fan, A. Argoti, R. Vidal-Michel and A. More, Stochastic modeling of thermal disinfection of bacteria according to the logistic law, *AIChE Journal* **51**(9), 2615-2618 (2005).
9. F. Marra and V. Romano, A mathematical model to study the influence of wireless temperature sensor during assessment of canned food sterilization, *Journal of Food Engineering* **59,** 245-252 (2003).
10. A. M. Van Loey, M. E. Hendrickx, S. De Cordt, T. Haentjens and P. Tobback, Quantitative evaluation of thermal processes using time temperature integrators, *Trends in Food Science & Technology* **7,** 16-26 (1996).
11. M. E. Hendrickx, Z. Weng, G. Maesmans and P. Tobback, Validation of time temperature integrator for thermal processing of food under pasteurisation conditions, *International of Food Science and Technology* **27,** 21-31 (1992).
12. P. S. Taoukis and T. P. Labuza, Reliability of time temperature indicators as food quality monitors under nonisothermal conditions, *Journal of Food Science* **54,** 789-792 (1989).

13. G. Tucker and S. D. Holdsworth, Mathematical modelling of sterilisation and cooking processes for heat preserved foods, applications of a new heat transfer model, *Trans I ChemE* **69,** 5-12(1991).

OPTIMAL DEFLECTION YOKE TUNING*

V. VAITKUS†
Process Control Department, Kaunas University of Technology, Studentų 48-327, Kaunas, 3000, Lithuania

A. GELŽINIS
Department of Applied Electronics, Kaunas University of Technology, Studentų 50-302, Kaunas, 3000, Lithuania

R. SIMUTIS
Process Control Department, Kaunas University of Technology, Studentų 48-327, Kaunas, 3000, Lithuania

A high quality deflection yoke (DY) is one of the most important factors for a high quality monitor. The role of the deflection yoke is to deflect electron beams in horizontal and vertical directions. If the magnetic field is formed incorrectly, misconvergence of beams may occur resulting in blurred image on the screen of the monitor. The magnetic field of DY may be corrected by sticking one or several ferroelastic shunts on the inside surface of the deflection yoke. Also some secondary balance parameters need to be controlled. Because of the complexity of the process it is not easy to determine how to stick the shunt. Therefore, two optimization methods were used for optimal tuning shunts position finding. This paper present research results and its application in industry.

1. Introduction

Still the most widely used display devise for television monitors is the color cathode ray tube (CRT).

The CRT produces visible light by bombardment of a thin layer of phosphor material by an energetic beam of electrons. Generally in the color CRT three electron guns produce three beams: red (R), green (G) and blue (B). The role of the deflection yoke (DY) is to deflect electron beams in horizontal and vertical directions. If the magnetic field of DY is formed incorrectly, misconvergence of beams may occur resulting in blurred image on the screen of the monitor. Small misconvergence can be eliminated by sticking one or several ferroelastic shunts

* This work is supported by Lithuanian State Science and Studies Foundation

on the inner part of the DY. The quality requirements for flat monitors are extremely high and requirements for DY also increase. When DY tuning is done manually by human expert, the quality of the tuning is determined by human operator experience. Because of DY manufactures strive for higher quality, the DY tuning process need to be automated in order to eliminate dependence on human experience

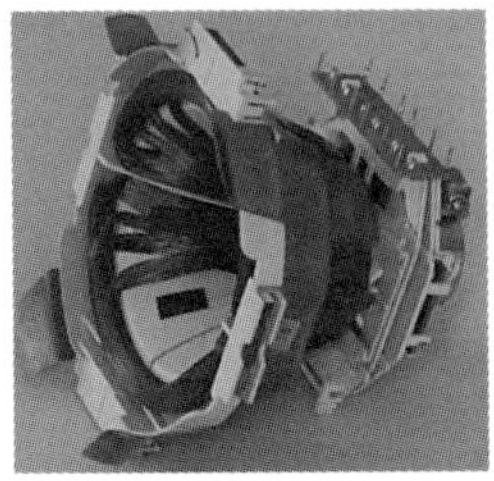

Figure 1. Deflection yoke.

2. What Was Done Before?

Heretofore some DY tuning strategies and methods were published.

Chung[1] introduced an intelligent adjustment process using the knowledge-base of the ferrite sheet for manufacturing. In the next paper he presented an intelligent knowledge-base controller with neuro-fuzzy modeling for deflection yoke's magnetic field control. Where approximately shunts positions were defined by fuzzy model and final positions were obtained by using gradient method. Once more decision support system based on fuzzy model was presented by Song *et al.*[2] But nowhere in the papers experimental results was presented or discussed. The authors published some works about DY tuning strategies and methods. But the main disadvantage of the methods proposed is that they were used to evaluate the residual misconvergence of the beams only between red (R) and blue (B) beams or only in 9 points on the screen. As it was mentioned the requirements for DY tuning quality highly increase and creation of decision support system became more complex.

3. Artificial Neural Networks

Artificial neural networks have been applied very successfully in the identification and control of dynamic systems.

The universal approximation capabilities of the multi-layer perceptron make it a popular choice for modeling nonlinear systems. So for evaluation of the shunt influence to beams misconvergence the multilayer perceptron was selected.

It has one hidden layer and output layer (Fig. 2). The perceptron network consist of d inputs, m hidden, c output neurons. It is feed-forward neural network with connections running from every neuron in one layer to every neuron in next layer through a set of weights w_{md} and w_{cm}, but with no other connections permitted. The network indices m and d indicate that w_{md} is the strength of the connection from the d-th input to the m-th neuron.

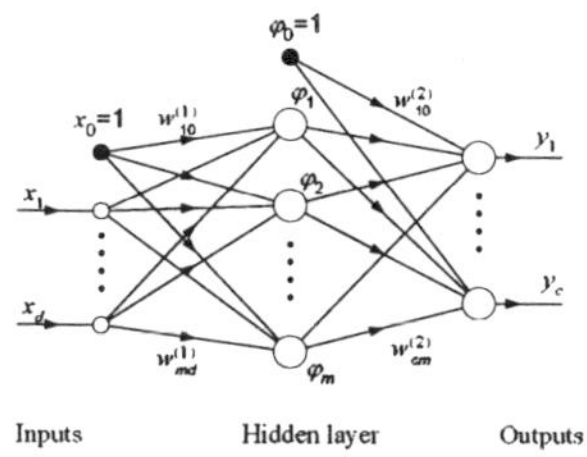

Figure 2. Structure of neural network.

We can write the analytic function corresponding to Fig. 3 as follows. The output of the j-th hidden neuron is obtained by first forming a weighted linear combination of the d input values and adding a bias, to give:

$$a_j = \sum_{i=1}^{d} w_{ji}^{(1)} x_i + w_{j0}^{(1)} \tag{1}$$

Here $w_{ji}^{(1)}$ denotes a weight in the first layer, going from input i to hidden neuron j, and $w_{j0}^{(1)}$ denotes the bias for hidden neuron j. The activation of hidden neuron j is then obtained by transforming the linear sum using an activation function g(·) to give:

$$\varphi_j = g(a_j) \tag{2}$$

In our case g(·) is a hyperbolic tangent transfer function. It calculates its output according to:

$$\varphi_{\mathrm{j}} = 2/(1 + \exp(-2 * \mathrm{a}_{\mathrm{j}})) - 1. \tag{3}$$

For each output neuron k (k = 1,...,c), we construct a linear combination of the outputs of the hidden neurons:

$$a_k = \sum_{j=1}^{m} w_{kj}^{(2)} \varphi_j + w_{k0}^{(2)} \tag{4}$$

Then neural network output is:

$$y_k = \tilde{g}(a_k) \tag{5}$$

In our case activation function $\tilde{g}$ is linear, and if we combine (1), (2), (4), (5) and absorb bias into the weights:

$$y_k = \sum_{j=0}^{m} w_{kj}^{(2)} g\left(\sum_{i=0}^{d} w_{ji}^{(1)} x_i\right) \tag{6}$$

4. Collecting of the Data for ANN Training

The misconvergence was evaluated in 16 positions of the monitor screen as shown in the Fig. 3.

1	2	3	4	5
6				7
8				9
10				11
12	13	14	15	16

Figure 3. Location of measuring points on the screen in our case.

It was done by measuring the beams misconvergence distances between blue and red *B-R*, red and green *R-G*, blue and green *B-G* beams according to x and y directions. In the earlier works it was assumed that the distance *R-B* is the most important for human operators during the DY tuning procedure but in practice this assumption is not always valid. Considering this presumption we have 96 primary parameters need to be controlled. For the preparation of the experimental measurement data eight DYs, type Philips 2180, with minimal initial misconvergence were used. The experimental results were received by sticking ferroelastic shunt and by measuring how much the beams misconvergence is changed between each one of three beams $\delta(B\text{-}R)$, $\delta(R\text{-}G)$ and $\delta(B\text{-}G)$. The position of the elastic shunt is given by distance d, as measured from the outermost border in depth of a DY, and an angle θ, as measured from the vertical axis. The shunt was sticked with different angle interval $\theta = 0° - 360°$ and distance $d = 0 - 40$ mm. During the experiment 5° and 5 mm steps were used to quantize the variables θ and d.

In order to improve the efficiency of decision support system we decided to use shunts with different size. The proportions for shunts were set as follows: 6x12mm – 0.75, 8x16mm – 1, 10x20mm – 1.25. To evaluate the shunt size influence to beams misconvergence in all angle (0,...,360)° and distance(0,…,40)mm interval, the different size shunts were placed in predetermined angle and distance positions. So we constructed neural network with three inputs (angle, distance and size) and three outputs *δ(B-R)*, *δ(R-G)* and *δ(B-G)* (Fig. 4).

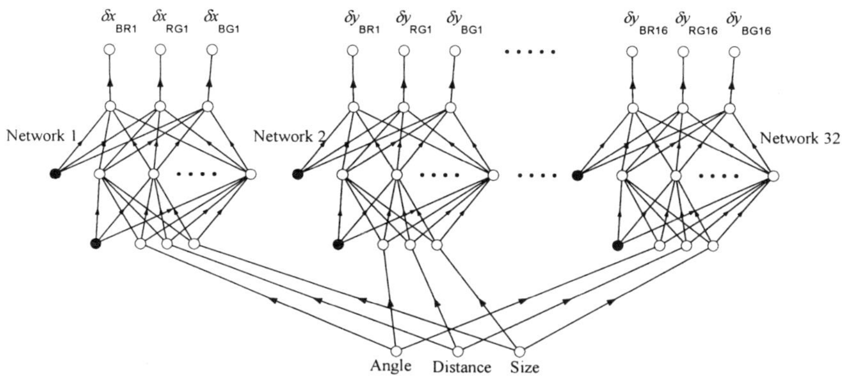

Figure 4. Structure of neural networks system for three inputs.

It has one hidden layer with 15 hidden neurons. The hyperbolic tangent transfer function type was selected for hidden layer neurons, and linear for output neurons. Levenberg-Marquardt training algorithm[3] was used for ANN training. Some simulation results are presented in Fig. 5. Neural network function depicted with continuous line.

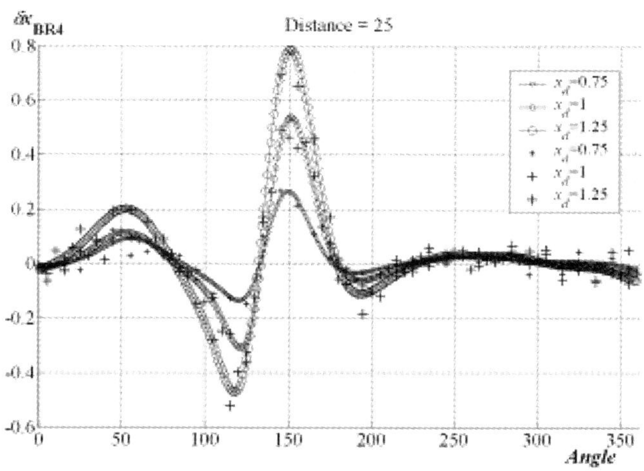

Figure 5. Different size shunts influences to beams BR misconvergence in measuring point 4 according X direction.

At the beginning of the DY tuning the geometrical parameters of the image on the screen is measured. If they are outside allowable interval DY orientation on cathode ray tube is changed by three step motors. Let's call the geometrical parameters of the image as DY balance parameters. Each one of the motors may correct one of balance parameters.

Experimental investigation showed that ferroelastic shunts have the sufficient influence to balance parameters. So after placing the shunt DY orientation on CRT will be changed again and beams misconvergence parameters will change also. To avoid this we need to predict shunt influence to balance parameters and then balance parameters correction influence to beams misconvergence if we want to have a desirable DY tuning result after shunts placing.

For the prediction of shunt influence to balance parameters neural network with 1 hidden layer was selected. Network structure is similar like presented in Fig. 4 and it consists of three networks.

During experimental investigation it was found that the dependence between balance parameters and beams misconvergence is linear. Considering that for the prediction of balance influence to beams misconvergence the second order polynomial equation was selected (7). Polynomial equation coefficients were estimated by Levenberg-Marquard algorithm.

$$\delta s_k = b_{0k} + \sum_{i=1}^{3} b_{ik} v_i + \sum_{i=1, j=i}^{3} b_{ijk} v_i v_j . \ \mathrm{k} = 1, \ldots, 96 \tag{7}$$

where b_k - polynomial equation coefficient of kth component, v_i – ith component of balance parameters vector V=(YVB, TBPINB, ROTATION).

5. Search Procedure

Since we can infer the relationship between input and output through the neural network, our decision support system consist of the 32 neural networks for primary parameters prediction, 3 neural networks for shunt influence to DY balance parameters, and polynomial equation system for the prediction of DY balance compensation influence to beams misconvergence (Fig. 6).

If misconvergence is occurred, a sheet is attached in initial position. From this position, the correct position is found by minimizing the error between the desired and simulated output vector. The optimal position pos_{opt} to place a correction shunt is given by:

$$pos_{opt} = \arg(\min_k(Krit_k)) \tag{8}$$

where $Krit_k$ – cost function which characterize DY parameters with k number of shunts placed on the DY. It's given by:

$$Krit_k = sum_{i=1}^{96}\left(\left(\frac{S_{ik}}{S_i^{\max}}\right)^{30}, \left(\frac{S_{ik}}{S_i^{\min}}\right)^{30}\right), \tag{9}$$

where S_{ik} – 96 residual misconvergence parameters. S_i^{min} – min allowable value of the ith parameter. S_i^{max} –max allowable value of the ith parameter.

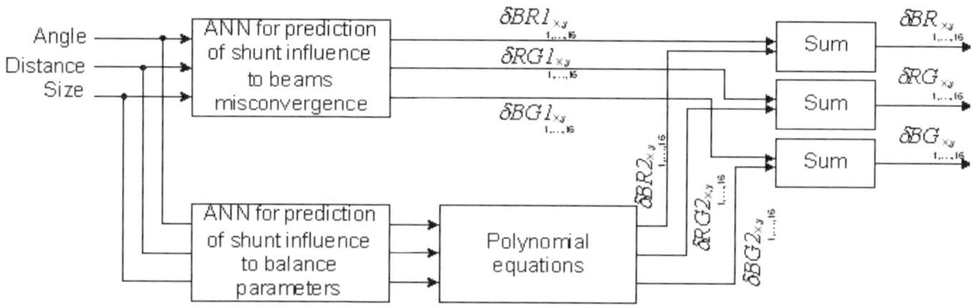

Figure 6. Structure of decision support system.

The parameter vector s_{ik} of DY with more than one correction shunt placed on DY is calculated as follows:

$$S_{ik} = s_0 + z_1 + \ldots + z_k, \quad k = 1,\ldots,N_s \tag{10}$$

where s_o is the parameter vector of DY without any correction shunt, z_k is the vector of changes of primary parameters (the neural network simulation result) caused by a correction shunt placed in the kth position. N_s – is the number of shunts used for DY tuning.

Our investigations have shown that the linearity assumption made in (10) approximately holds in practice. It is assumed that the deflection yoke is successfully tuned if the value of the criterion is less than 1.

Initial shunts positions are obtained by stochastic *simulated annealing SA* algorithm. *SA* procedure is a random search process and it's effective for a globally optimal point search.[4] The proposed *simulated annealing* algorithm can be realized as follows:

1. Randomly select the initial parameter vector x_0 (shunt position). Choose "Temperature" T value such that exp(-ΔE/T) ≥0.999 for all error E changes ΔE.
2. Select a component xi of x uniformly at random.
3. Change vector w to vector x' such that $x'_i=\xi_i$, where ξ_i – the value chosen uniformly at random from the search area X_i.
4. Set

$$x^{new} = \begin{cases} \mathbf{w}' & \text{with probability } \exp\left[\dfrac{-\left[E(x) - E(x)\right]^{+}}{T}\right]; \\ \mathbf{w} & \text{else.} \end{cases} \qquad (11)$$

5. . If it was made M successful changes of vector x (changes for which the value of E dropped) or N total changes in w have occurred since the last change in temperature, then set the value T to βT. M is typically an order of magnitude smaller than N and β is typically between 0.8 and 0.9999.
6. If the minimum value of E has not decreased more than ε (a small constant) in last S (S>>N) iteration then stop the search. Otherwise go to step 2.

The final shunt position search was executed by *gradient descent*. Let's take an example and notations shown in Fig. 7.

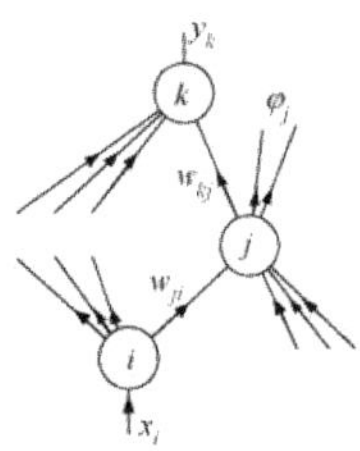

Figure 7. ANN structure and accepted notations.

From (8) and (9) the cost function component e_k:

$$e_k = \left(\frac{s_i + z_k}{s_k^{\max,\min}} \right)^{30}, \qquad (12)$$

Partial derivative of the cost function with the respect to the input to network could be found as follow:

$$\frac{\partial e_k}{\partial x_i} = \sum_j \frac{\partial e_k}{\partial a_j} \frac{\partial a_j}{\partial x_i} \qquad (13)$$

If $\delta_j = \frac{\partial e_k}{\partial a_j}$, and $\frac{\partial a_j}{\partial x_i} = w_{ji}$, then (13) could be expressed as follow:

$$\frac{\partial e_k}{\partial x_i} = \sum_j \delta_j w_{ji} \tag{14}$$

Therefore, after some equations rearrangement the final expression of δ_j :

$$\delta_j \equiv \frac{\partial e_k}{\partial a_j} = \sum_k \frac{\partial e_k}{\partial a_k}\frac{\partial a_k}{\partial a_j} = g'(a_j)\sum_k w_{kj}\delta_k = \left(1 - g^2(a_j)\right)\sum_k w_{kj}\delta_k\ , \tag{15}$$

where δ_k for output neurons:

$$\delta_k \equiv \frac{\partial e_k}{\partial a_k} = \frac{\partial y_k}{\partial a_k}\frac{\partial e_k}{\partial y_k} = \tilde{g}\left(a_k\right)\frac{\partial e_k}{\partial y_k} = \frac{30\left(s_k + z_k\right)^{29}}{\left(s_k^{\max,\min}\right)^{29}} \tag{16}$$

In order to accelerate the shunts positions search the RPROP ("resilient backpropagation") algorithm[5] for changing of each one of the input vector component was used. Partial derivative of the cost function with respect to input was used only to define the descent direction. The learning rate η_i in (17) was changed (increased or decreased) by 20% depending on successful or not the last step was. The start value of learning rate was set 0.001.

$$\Delta x_i = -\eta_i \operatorname{sgn}\left|\frac{\partial E}{\partial x_i}\right| \tag{17}$$

where Δx_i – input vector ith component, η_i – learning rate for ith input vector component.

6. Experiments

The decision support system proposed was tested with 253 deflection yokes.

Maximum allowable number of shunts for DY tuning is 8. For each one of deflection yoke 20 search starts executed, 10000 SA and 100 gradient descent iterations for each start selected. In order to check the advantage of different shunt size using two decision support systems were tested. The first system used fixed shunt size (8x16) mm, second used three different shunt sizes. The test results are presented in Table 1.

Table 1. Decision support system test results.

Decision Support System Type	The average of shunts used	The average of *Krit* received
Fixed shunt size	2.62	0.841
Three different shunts sizes	2.32	0.833

The *Krit* presented in Table 1 was calculated by equation:

$$Krit_{\max} = \max_{i}^{16}\left\{\max_{j}^{3}\left[\max_{k}^{2}\left(\frac{S_{ijk}}{S_{ijk}^{\max,\min}}\right)\right]\right\}, \tag{18}$$

where i – number of measuring points on the screen, j – number of distances between beams (*BR, RG, BG*), k – x and y directions.

It's clear that decision support system with different shunt size have more advantages. It used less shunts and the criterion received was smaller. These advantages became more observable when for DY tuning 4 or more shunts were used. I took about 3 seconds to calculate positions for three shunts. Based on the result obtained we recommend to use the decision support system proposed in real DY adjustment process.

Acknowledgment

The technical and financial support from Join Stock Company "Vilniaus Vingis" and Lithuanian State Science and Studies Foundation is gratefully acknowledged.

References

1. B. M. Chung, C. O. Jung and J. S. Yun, Quality control system for deflection yoke of CDT, *Proc. 21st International Conference on Industrial Electronics, Control and Instrumentation*, 1543-1548 (1995).
2. W. K. Song *et al.*, Convergence adjustment of deflection yoke using soft computing techniques, *Proc. IEEE International Conference on Fuzzy Systems*, vol. 1, 1543-1548 (1995).
3. J. J. More, The Levenberg-Marquardt algorithm: implementation and theory, In: G. A. Watson (ed.), *Numerical Analysis, Lecture Notes in Mathematics* **630**, Springer Verlag, 105-116 (1977).
4. S. Kirkpatrick, C. D. Gelatt and M. P. Vecchi, Optimization by simulated annealing, *Science* **220**, 671-680 (1983).
5. M. Riedmiller and H. Braun, A direct adaptive method for faster backpropagation learning: the RPROP algorithm, *Proceedings of the IEEE International Conference on Neural Networks (ICNN)*, 586-591 (1993).

ANALYSIS OF AN EXTRACTIVE FERMENTATION PROCESS FOR ETHANOL PRODUCTION USING A RIGOROUS MODEL AND A SHORT-CUT METHOD

O. J. SÁNCHEZ

Department of Chemical Engineering, University College London, Torrington Place London WC1E 7JE, United Kingdom
Department of Engineering, University of Caldas Calle 65 № 26-10, Manizales, Colombia

L. F. GUTIÉRREZ

Department of Chemical Engineering, National University of Colombia at Manizales Cra. 27 № 64-60, Manizales, Colombia

C. A. CARDONA[†]

Department of Chemical Engineering, National University of Colombia at Manizales Cra. 27 № 64-60, Manizales, Colombia

E. S. FRAGA

Department of Chemical Engineering, University College London, Torrington Place London WC1E 7JE, United Kingdom

Extractive fermentation is based on the removal of inhibitory compounds from the culture broth by an extractive agent allowing the process intensification. In this work, a rigorous approach for the description of extractive fermentation for ethanol production was utilized. With this aim, fermentation kinetics models were coupled with models describing liquid-liquid equilibrium in order to simulate the continuous culture. A short-cut method based on the principles of thermodynamic-topological analysis is proposed for studying the behaviour of the process. The feasibility of different sets of operating parameters was examined. Using the mentioned tools, a general strategy of optimization was formulated.

1. Introduction

Continuous fermentation is the cultivation regime that offers higher productivities. Ethanol fermentation is one of the cases where continuous culture has been implemented at industrial scale. Unfortunately, natural regulation

[†] Corresponding author. E-mail: ccardonaal@unal.edu.co

mechanisms of microbial biomass lead to reduced production of ethanol due to the inhibition effect of products present in the culture broth.[1] A reasonable approach for increasing the productivity of alcoholic fermentation is the removal of the product that causes the inhibition through an extractive biocompatible agent (solvent) that favours the migration of ethanol to solvent phase, a process known as extractive fermentation. As a solvent, *n*-dodecanol has been chose by its very low toxicity for ethanol-producing microorganisms and its ethanol selectivity.[2] A model for describing the extractive fermentation process for continuous production of ethanol from glucose-containing medium was reported;[3] in this model, a simple relationship between ethanol concentration in aqueous phase and ethanol content in solvent phase was considered. Furthermore, the kinetic description of microbial growth did not consider the inhibition effect due to high substrate concentrations. Lignocellulosic biomass is a promising feedstock for ethanol production, but it is necessary to overcome the technological difficulties limiting its widespread utilization. New alternatives for the intensification of this process such biomass cofermentation using genetically modified strains have been proposed.[4] In a previous work,[5] these features were considered for the description of extractive fermentation for batch and fed-batch regimes. The objective of this work was to model the extractive fermentation for ethanol production from biomass using a rigorous mathematical description that couples both kinetic and extraction phenomena. In addition, a short-cut approach for analyzing this process was proposed, as well as an overall strategy of optimization.

2. Rigorous Modelling of Continuous Extractive Fermentation

To describe the continuous process of extractive fermentation for fuel ethanol production, *n*-dodecanol was selected as an extracting agent. Feed aqueous stream containing nutritive components is added to a CSTR, where a solvent stream is continuously fed as well. Fed sugars are generated during the pretreatment of lignocellulosic biomass in which major polysaccharides are broken down into elementary sugars like hexoses (glucose) and pentoses (mainly xylose). Formed sugars are converted into ethanol in the reactor. Ethanol is distributed between aqueous and organic (solvent) phases, diminishing its concentration in the culture aqueous broth and allowing the reduction of product inhibition effect on the microorganisms. Ethanol-enriched solvent phase is continuously removed from the reactor through a decanting unit. This stream is sent to a flash unit in order to recover the obtained ethanol and to regenerate the solvent, which can be recycled to the CSTR.

To develop a rigorous model describing both the fermentation and liquid-liquid extraction processes, the cultivation kinetics is coupled with an extraction model. The liquid-liquid equilibrium was described by an algorithm that employs the UNIFAC equations for the calculation of activity coefficients of components in each phase. This algorithm was integrated into the *ModELL* software, which was designed by our research group and that couples two convergence algorithms (Newton-Raphson and False Position Method) in order to calculate the liquid fraction of each phase. *ModELL* was developed in Delphi package version 7.0 (Borland Software Corporation, USA). The kinetic model of fermentation was taken from Leksawasdi *et al.*[6] It describes the simultaneous consumption by a recombinant strain of *Zymomonas mobilis* of two main substrates contained in the lignocellulosic hydrolyzates: glucose and xylose. This model takes into account the biomass growth and ethanol production and considers substrate limitation, substrate inhibition and ethanol inhibition.

The following assumptions were considered during the formulation of overall model of extractive fermentation: a) the substrate uptake, biomass formation and product biosynthesis are carried out only in the aqueous phase; b) ethanol is the component migrating to the solvent phase and small amounts of water can migrate to the organic phase depending of the solvent; c) solvent is biocompatible with the microorganisms and does not have any effect on the fermentation; d) stirring in the bioreactor ensures total mixing between liquid phases and does not produce damage to the cells. The configuration corresponding to continuous extractive fermentation involves the feeding of culture medium and solvent to the reactor and the continuous removal of both aqueous and solvent phases from the reactor in a separate way (see Fig. 1).

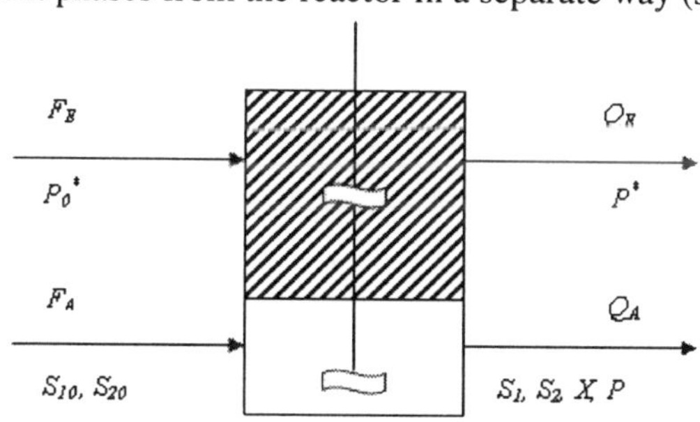

Figure 1. Schematic diagram of continuous extractive fermentation for ethanol production. The decanting unit is not shown.

In this case, the flow rate of influent aqueous stream (F_A) is greater than the flow rate of effluent aqueous stream (Q_A) because of the migration of ethanol to the solvent phase. Mass balance equations representing this process are as follows:

$$-Q_A X + V_A r_X = 0 \tag{1}$$

$$F_A S_{10} - Q_A S_1 - V_A r_{S1} = 0 \tag{2}$$

$$F_A S_{20} - Q_A S_2 - V_A r_{S2} = 0 \tag{3}$$

$$F_E P_0^* - Q_A P - Q_E P^* + V_A r_P = 0 \tag{4}$$

$$F_A + F_E - Q_A - Q_E = 0 \tag{5}$$

$$P^* = k_{EtOH} P, \tag{6}$$

where r_X is the cell gowth rate (in g.L^{-1}.h^{-1}), r_{S1} and r_{S2} are the glucose and xylose consumption rates, respectively (in g.L^{-1}.h^{-1}), and r_P is the ethanol formation rate (in g.L^{-1}.h^{-1}); X, S_1, S_2 and P are the concentrations of cell biomass, glucose, xylose and ethanol in the aqueous effluent from the bioreactor (in g.L^{-1}), and X_0, S_{10}, S_{20} and P_0 are the corresponding concentrations in the aqueous feed stream (in g.L^{-1}); P^* and P_0^* are the ethanol concentrations in the solvent effluent and in the solvent feed streams, respectively (g.L^{-1}). For solving the system of equations, a constant solvent volume/aqueous volume ratio is assumed. The ethanol concentration in the solvent phase (P^*) that is in equilibrium with ethanol concentration in the aqueous phase is determined using the distribution coefficient k_{EtOH}, as shown in (6), which is calculated by the algorithm for liquid-liquid equilibrium. The determination of all variables involved in the model is performed using the software *ModELL*. For specified inlet aqueous dilution rate ($D_{Ai} = F_A / V_A$) and solvent feed flow rate/aqueous feed flow rate ratio ($R = F_E / F_A$), the program requires concentrations of cell biomass, substrates and ethanol in feed streams.

3. Preliminary Short-Cut Method

In order to develop a short-cut method for extractive fermentation process, the principles of thermodynamic-topological analysis[7] were applied to studied conditions. In particular, the main components (substrate-water-product-solvent) can be represented in a quaternary diagram in order to locate initial conditions. For the representation of the reaction trajectory, and considering that the overall fermentation process is irreversible, a stoichiometric approach was utilized. Therefore, the fermentation is described as follows:

$$\underset{\text{Glucose}}{C_6H_{12}O_6} \xrightarrow{\textit{Z. mobilis}} \underset{\text{Ethanol}}{2C_2H_5OH} + 2CO_2$$

$$\underset{\text{Xylose}}{3C_5H_{10}O_5} \xrightarrow{\textit{Z. mobilis}} \underset{\text{Ethanol}}{5C_2H_5OH} + 5CO_2$$

Once determined the proportion between substrate (expressed as the sum of both sugars) and ethanol considering a maximum stoichiometric yield from substrate of 0.511 $g.g^{-1}$, the inlet conditions are located in the ternary diagram water-ethanol-*n*-dodecanol, where the liquid-liquid equilibrium is represented as well. Finally, the balance lines corresponding to the operating conditions for the steady states were drawn in order to define the zone where the performance of extractive fermentation process is more stable and advantageous.

4. Results and Discussion

Based on the model of Leksawasdi *et al.*[6], the simulation of alcoholic extractive fermentation from biomass was performed with a glucose concentration of 100 $g.L^{-1}$ and a xylose concentration of 50 $g.L^{-1}$ in the feed aqueous stream. These concentrations correspond to those of lignocellulosic hydrolyzates. The behaviour of this process for R=2 in dependence of inlet dilution rate (D_{Ai}) is shown in Fig. 2. The cell wash-out occurs at dilution rates near 0.33 h^{-1}. These results were obtained for a solvent volume/aqueous volume ratio (V_E/V_A) of 2. Higher ethanol productivities are found in the range 0.25-0.30 h^{-1}. In order to elucidate the best operating value of D_{Ai}, GAMS software (General Algebraic Modeling System, GAMS Development Corporation, USA) was used for maximizing total ethanol productivity. For this, a simple linear relationship for (6) was considered. The optimal D_{Ai} was of 0.265 h-1.

The coupled algorithm was used for process simulation varying the solvent feed flow rate/aqueous feed flow rate ratio (R) for an inlet dilution rate of 0.265 h^{-1}. Best results were obtained for values greater than 4 that correspond to increased amount of consumed substrates. Both total productivity and productivity for ethanol recovered from solvent phase are approaching to constant values. For R higher than 8, the model predicts the formation of homogeneous mixture without extraction. The simulation of this process modifying both R and D_{Ai} shows that the zone of manipulating variables with higher productivities (inlet concentrations of glucose and xylose in the aqueous stream of 100 $g.L^{-1}$ and 50 $g.L^{-1}$, respectively) corresponded to dilution rates near wash-out conditions, and to higher values of R within the range 1.29-7.9.

If the concentration of both substrates in the feed stream is varied, the problem becomes more complex. Physically, the increase in substrate can be achieved as a result of evaporation of initial hydrolyzate obtained from biomass pretreatment. For this reason, the proportion of glucose and xylose in the inlet aqueous stream should be constant and equal to 2:1. Best values of total productivity and ethanol productivity recovered from solvent phase correspond to an inlet concentration of total sugars of about 600 g.L^{-1}. The simulation was carried out until the concentration of sugars was less than or equal to 600 g.L^{-1}, which corresponds to maximum solubility of these sugars in water.

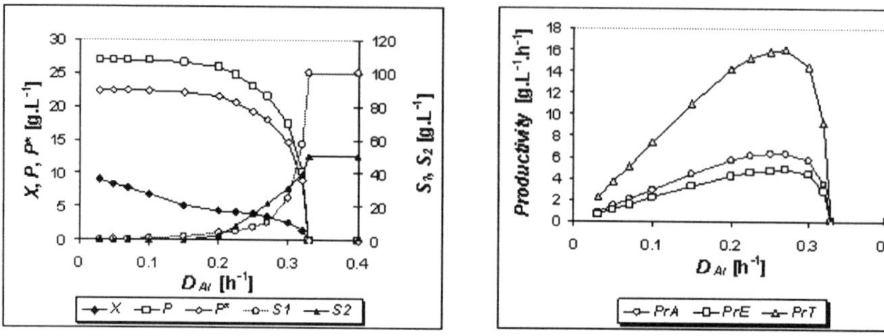

(a) (b)

Figure 2. Continuous extractive fermentation using *n*-dodecanol. Effect of inlet aqueous dilution rate (D_{Ai}) on: (a) effluent concentrations of glucose (S_1), xylose (S_2), cells (X), ethanol in aqueous phase (P), and ethanol in solvent phase (P^*); (b) total ethanol productivity (PrT), productivity for ethanol recovered from aqueous phase (PrA), and productivity for ethanol recovered from solvent phase (PrE). Concentration of sugars in feed aqueous stream: glucose, 100 g.L^{-1}; xylose, 50 g.L^{-1}.

The procedure for locating the steady states in the concentration simplex using a short-cut approach is illustrated in Fig. 3. The process is ideally divided into two steps: microbial conversion and liquid-liquid extraction. The initial sugar concentration is represented in the quaternary diagram by the point *A* (see Fig. 3a). The transformation of sugars into ethanol is shown by the line *AB*, being *B* the state of the system where the total amount of produced ethanol is represented. This point is the starting mixture for the liquid-liquid equilibrium. The line *BC* represents the addition of *n*-dodecanol to the aqueous medium containing ethanol (Fig. 3b). This line lies in the ternary diagram water-ethanol-*n*-dodecanol, where the zone of heterogeneous mixtures is drawn too. Vertical lines represent the geometric place of points that represent the operating conditions related to the solvent feed stream/aqueous feed stream (*R*). The intersection of these vertical lines with the line *BC* (point *D*) represents the theoretical conditions corresponding to the mixtures before the separation in phases (the equivalent to the feed mixture in a liquid extractor). Through tie lines, the compositions of the extract (*E*) and the raffinate (*W*) are obtained.

These correspond to the compositions of solvent phase effluent and aqueous phase effluent, respectively. For identical inlet concentrations of sugars in the aqueous stream, the position of the starting point *B* changes when inlet dilution rate varies. For example, if D_{Ai} increases, the new line *B'C* will lie below the original line *BC*, which it is explained by a major dilution of ethanol and, therefore, the line approaches to the bottom edge of the ternary diagram.

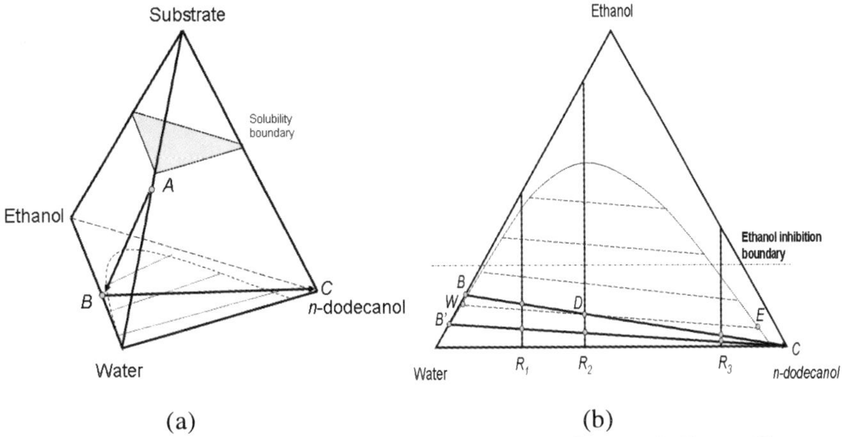

Figure 3. Representation of extractive fermentation. (a) Quaternary diagram. (b) Ternary Diagram.

When the concentration of sugars in the feed aqueous stream increases up to the limit imposed by the maximum solubility of the sugars in water (see Fig. 3a), the total amount of ethanol formed, based on the stoichiometry, increases as well, and therefore, the point *B* is displaced towards ethanol in the edge of the ternary diagram corresponding to the water-ethanol mixtures (see Fig. 4). This new point *B"* should be below the line corresponding to the substrate solubility boundary determined by the point *H*. For simplicity, this boundary is represented by a plane in Fig. 3a. The ethanol concentration in aqueous phase should not be greater than ethanol concentration above which cell growth rate is inhibited. This ethanol inhibition boundary is determined by the point *I*. Therefore, the feasible operation zone for extractive fermentation corresponds to the shaded area delimited by the points R_2D' and the binodal curve in the Fig. 4. When initial substrate concentrations in the feed stream are changed, the vertical lines showing the values of *R* are displaced in comparison to the lines corresponding to initial point *D* in Fig. 3b.

Using this short-cut approach, the zone of feasible operating points can be easily determined. Let us analyze the extreme case when the feed aqueous stream has the maximum allowable concentration of sugars. From the fermentation stoichiometry, this condition corresponds to an inlet concentration in the feed aqueous stream of about 600 g.L^{-1} of total sugars. Assuming a 95% yield, the total amount of ethanol that could be produced is 0.486 g.g^{-1}, that implies a theoretical starting ethanol concentration of 291.6 g.L^{-1} (approximately, an ethanol mass fraction of 0.42). This value determines the position of the substrate solubility boundary (point *H* in Fig. 4). Since the concentration of ethanol in the aqueous phase (raffinate) should not be above the ethanol inhibition boundary (approximately 10% w/w), the operation conditions represented by the line R_3 should be such that the ethanol content in the raffinate corresponding to point *D"* be equal to the ethanol content of the point *I* to avoid product inhibition. In this way, the area delimited by the points $R_3D"EK$ is the zone of feasible steady states for given conditions of the process with the maximum concentrations of sugars in the culture broth. For a given inlet dilution rate of 0.1 h^{-1}, an *R* ratio of 3.6 (line R_2), and a working volume of 1 L, the location of the point *D'* and the corresponding composition of the extract, can be found. In this case, the ethanol mass content of the extract and raffinate is of 7.5% and 9.0%, respectively, resulting in a total ethanol productivity of 26.4 g.L^{-1}.h^{-1}. The productivity calculated by the rigorous model using *ModELL* is 28.86 g.L^{-1}.h^{-1}. Hence, short-cut method allowed determining the feasibility of operating parameters and an estimation of the productivities.

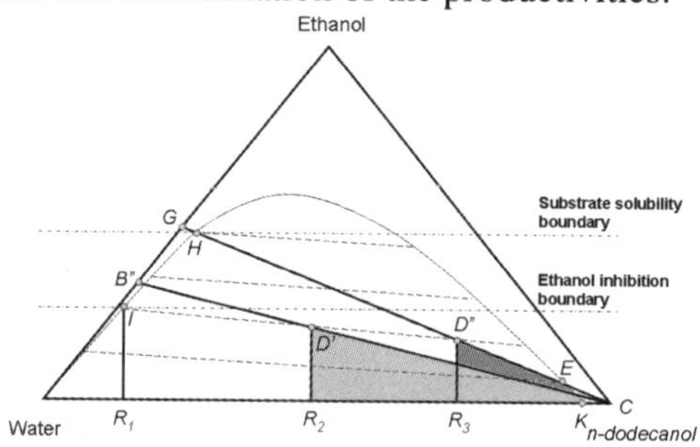

Figure 4. Representation of extractive fermentation process for different concentrations of sugars in the inlet aqueous streams.

The delimited zones can be taken into account for the development of a preliminary strategy of optimization. Since the region of feasible steady states was determined in the ternary diagram (see Fig. 4), the values range of such manipulating variables as inlet dilution rate, the *R* ratio, and the concentration of the sugars in the inlet aqueous stream are known and can be bounded for solving an optimization algorithm. The GAMS system was utilized to find the optimal value of above-mentioned variables that maximize the total ethanol productivity using the NLP solver CONOPT3. With this aim, liquid-liquid equilibrium relationships were simplified for generating a way to evaluate the ethanol concentration in both phases during extractive fermentation. For this, the distribution coefficient k_{EtOH} was assumed to be linearly dependent of the total concentration of substrates. A good concordance with the data obtained from the *ModELL* was achieved, especially for high values of substrates concentration. The results of this optimization are presented in Table 1. From this table, it is evident that calculated optimal variables are indeed in the zone predicted by the short-cut method, and the predicted increase in total productivity is effectively attained, as it is shown by the rigorous model.

Table 1. Preliminary optimal results for manipulating variables calculated by GAMS and corresponding values calculated by ModELL software.

Variable	R	D_{Ai} [h^{-1}]	S_{10} [$g.L^{-1}$]	S_{20} [$g.L^{-1}$]	PrT [$g.L^{-1}.h^{-1}$]	P [$g.L^{-1}$]	P^* [$g.L^{-1}$]
GAMS	3.038	0.185	400	200	51.46	40.52	73.55
ModELL	3.038	0.185	400	200	54.79	40.31	73.88

For a more accurate solution of the optimization problem, the rigorous description of the equilibrium model should be coupled or embedded into the GAMS code. Further analysis of this extractive fermentation process could include the formulation of an objective function that considers, besides ethanol productivity, other performance indexes like the conversion of sugars (better utilization of the feedstock) or the amount of generated wastewater (evaluation of environmental impact). These issues will be developed in upcoming works.

Once obtained a global picture of the space of operating conditions and their optimal values for the studied process, experimental runs should be performed in order to confirm the validity of the given theoretical approach. In this manner, the acquired insight of the process will make possible the reduction of expensive experimental work in the search of the optimal operation. Experimental data for alcoholic extractive fermentation using *n*-dodecanol available in the open literature do not allow the comparison with the present results mainly due to the fact that the reported operating conditions (immobilized cells, batch regime, coupled systems using a separate extractor-decanting unit) were different to the proposed in this work; for this reason, the needed experimental runs should be undertaken in the future.

5. Conclusions

The removal of valuable products from culture broths is a promising technology for the intensification of fermentation processes. The selection of proper solvent is a key aspect to develop this kind of processes successfully. On the other hand, the study of the behaviour of extractive fermentation can provide useful tools for defining the best operating parameters and suitable regimes in order to increase techno-economical indexes of biotechnological transformations. The proposed short-cut method based on the principles of thermodynamic-topological analysis allows getting a preliminary idea for approaching to the rigorous simulation. This approach makes possible the decrease in calculation time and in the number of experimental runs. Moreover, it helps to determine which data are required and the space of initial conditions where experimental efforts should be focused. The outcomes obtained in this work demonstrated the usefulness and advantages of this methodology when multivariate optimization is needed for the determination of the best operating parameters in such a complex process as the extractive fermentation.

Acknowledgements

The authors gratefully acknowledge the support provided by British Council for attendance at the INYS Workshop and the access to computational resources provided by University College London. The financial support of the Colombian Institute for Development of Science and Technology (Colciencias) and of the National University of Colombia at Manizales is also acknowledged.

References

1. J. J. Malinowski, *Biotechn. Advances* **19**, 525 (2001).
2. M. Gyamerah and J. Glover, *J. Chem. Technol. Biotech.* **67,** 145 (1996).
3. F. Kollerup and A. J. Daugulis, *Biotech. and Bioeng.* *27,* 1335 (1985).
4. J. R. Mielenz, *Current Opin. in Microbiol.* **4,** 324 (2001).
5. L. F. Gutiérrez, O. J. Sánchez and C. A. Cardona, In: *PRES'05*. Giardini Naxos, Italy (2005).
6. N. Leksawasdi, E. L. Joachimsthal and P. L. Rogers, *Biotechn. Lett.* **23,** 1087 (2001).
7. Yu. A. Pisarenko, L. A. Serafimov, C. A. Cardona, D. E. Efremov and A. S. Schuwalov, *Rev. in Chem. Eng.* **17(4)**, 253 (2001).

APPLICATION OF GENERIC MODEL CONTROL FOR AUTOTROPHIC BIOMASS SPECIFIC GROWTH CONTROL

J. REPŠYTĖ AND R. SIMUTIS

Process Control Department, Kaunas University of Technology, Studentu st, 48-327 LT-51367 Kaunas, Lithuania

A model based control approach, known as Generic Model Control (GMC), was analyzed and proposed to the regulation of the specific growth rate of autotrophic biomass in wastewater treatment plant (WWTP). In GMC theory, the nonlinear process model is directly embedded in the control law. One of the most attractive features is that this control scheme solves an optimization problem in only one step of calculation. In aid of the complex WWTP simulator, economy efficient was analyzed if autotrophic biomass specific growth rate set point is increased at night and decreased at day time (night and day electrical energy tariffs).

1. A Short Introduction to Wastewater Treatment Process

Modern wastewater treatment is a fairly complex process, which includes several treatment steps before the wastewater is cleaned. A very typical strategy is to have four different process steps:[1]

- mechanical treatment;
- biological treatment;
- sludge treatment;
- chemical treatment.

2. Mathematical Model

The dynamic models are valuable tools for the plant operator or the designer in forecasting or explaining the performance of the wastewater treatment plant. They also can help to evaluate the efficiency of different process control strategies. The mathematical models on the basis of material balances are created for each wastewater treatment process. The mathematical models of some processes are not presented in this paper because they were already analyzed in earlier paper.[2]

2.1. *Biological Treatment*

The process is a completely mixed bioreactor with ten components.

The dynamic behaviour of the heterotrophic biomass concentration is affected by three different processes – aerobic growth, anoxic growth and decay – according to

$$\frac{dX_{B,H}}{dt} = \hat{\mu}_H (\frac{S_S}{K_S + S_S})(\frac{S_O}{K_{O,H} + S_O}) X_{B,H} + \hat{\mu}_H (\frac{S_S}{K_S + S_S}) \times \\ \times (\frac{K_{O,H}}{K_{O,H} + S_O})(\frac{S_{NO}}{K_{NO} + S_{NO}}) \eta_g X_{B,H} - b_H X_{B,H} \tag{1}$$

The situation for the autotrophic biomass concentration is simpler since the autotrophs do not grow in an anoxic environment. Consequently,

$$\frac{dX_{B,A}}{dt} = \hat{\mu}_A (\frac{S_{NH}}{K_{NH} + S_{NH}})(\frac{S_O}{K_{O,A} + S_O}) X_{B,A} - b_A X_{B,A} \tag{2}$$

The concentration of readily biodegradable substrate is reduced by the growth of heterotrophic bacteria (in both aerobic and anoxic conditions) and is increased by hydrolysis of slowly biodegradable substrate and the differential equation describing this is

$$\frac{dS_S}{dt} = -\frac{1}{Y_H} \hat{\mu}_H (\frac{S_S}{K_S + S_S})(\frac{S_O}{K_{O.H} + S_O}) X_{B.H} - \frac{1}{Y_H} \hat{\mu}_H (\frac{S_S}{K_S + S_S})(\frac{K_{O.H}}{K_{O.H} + S_O})(\frac{S_{NO}}{K_{NO} + S_{NO}}) \eta_g \times \\ \times X_{B.H} + k_h \frac{X_S / X_{B.H}}{K_X + X_S / X_{B.H}} ((\frac{S_O}{K_{O.H} + S_O}) + \eta_h (\frac{K_{O.H}}{K_{O.H} + S_O})(\frac{S_{NO}}{K_{NO} + S_{NO}})) X_{B.H} \tag{3}$$

The concentration of particulate organic nitrogen is increased by biomass decay and decreased by the hydrolysis process. The differential equation becomes

$$\frac{dX_{ND}}{dt} = (i_{XB} - f_p i_{XP})(b_H X_{B.H} + b_A X_{B.A}) - k_h \frac{X_{ND} / X_{B.H}}{K_X + X_S / X_{B.H}} \times \\ \times ((\frac{S_O}{K_{O.H} + S_O}) + \eta_h (\frac{S_O}{K_{O.H} + S_O})(\frac{S_{NO}}{K_{NO} + S_{NO}})) X_{B.H} \tag{4}$$

The concentration of soluble organic nitrogen is affected by ammonification and hydrolysis, according to

$$\frac{dS_{ND}}{dt} = -k_a S_{ND} X_{B,H} + k_h \frac{X_{ND} / X_{B,H}}{K_X + X_S / X_{B,H}} ((\frac{S_O}{K_{O,H} + S_O}) + \\ + \eta_h (\frac{S_O}{K_{O,H} + S_O})(\frac{S_{NO}}{K_{NO} + S_{NO}})) X_{B,H} \tag{5}$$

The concentration of slowly biodegradable substrate is increased by the recycling of dead bacteria according to the death-regeneration hypothesis and decreased by the hydrolysis process according to

$$\frac{dX_S}{dt} = (1 - f_p)(b_H X_{B,H} + b_A X_{B,A}) - k_h \frac{X_S / X_{B,H}}{K_X + X_S / X_{B,H}}((\frac{S_O}{K_{O,H} + S_O}) + \\ + \eta_h (\frac{K_{O,H}}{K_{O,H} + S_O})(\frac{S_{NO}}{K_{NO} + S_{NO}}))X_{B,H} \tag{6}$$

The shortest model equation is the one describing the concentration of inert particulate products arising from biomass decay, which is simply

$$\frac{dX_P}{dt} = f_p (b_H X_{B,H} + b_A X_{B,A}) \tag{7}$$

The ammonia concentration is affected by growth of all microorganisms as ammonia is used as the nitrogen source for incorporation into the cell mass. The concentration is also decreased by the nitrification process and increased as a result of ammonification of soluble organic nitrogen. This leads to a complex differential equation formulated as

$$\frac{dS_{NH}}{dt} = -i_{XB} \hat{\mu}_H (\frac{S_S}{K_S + S_S})(\frac{S_O}{K_{O.H} + S_O}) X_{B.H} - i_{XB} \hat{\mu}_H (\frac{S_S}{K_S + S_S})(\frac{K_{O.H}}{K_{O.H} + S_O})(\frac{S_{NO}}{K_{NO} + S_{NO}}) \times \\ \times \eta_g X_{B.H} - (i_{XB} + \frac{1}{Y_A}) \hat{\mu}_A (\frac{S_{NH}}{K_{NH} + S_{NH}})(\frac{S_O}{K_{O.A} + S_O}) X_{B.A} + k_a S_{ND} X_{B.H} \tag{8}$$

The concentration of nitrate is only involved in two processes– it is increased by nitrification and decreased by de-nitrification. The dynamic equation describing this is formulated below.

$$\frac{dS_{NO}}{dt} = -\frac{1 - Y_H}{2.86 Y_H} \hat{\mu}_H (\frac{S_S}{K_S + S_S})(\frac{S_O}{K_{O.H} + S_O}) X_{B.H} - \frac{1}{Y_H} \hat{\mu}_H (\frac{S_S}{K_S + S_S})(\frac{K_{O.H}}{K_{O.H} + S_O}) \times \\ \times (\frac{S_{NO}}{K_{NO} + S_{NO}}) \eta_g X_{B.H} + \frac{1}{Y_A} \hat{\mu}_A (\frac{S_{NH}}{K_{NH} + S_{NH}})(\frac{S_O}{K_{O.H} + S_O}) X_{B.A} \tag{9}$$

Finally, the oxygen concentration in the wastewater is reduced by the aerobic growth of heterotrophic and autotrophic biomass, according to

$$\frac{dS_O}{dt} = -\frac{1 - Y_H}{Y_H} \hat{\mu}_H (\frac{S_S}{K_S + S_S})(\frac{S_O}{K_{O,H} + S_O}) X_{B,H} - \frac{4.57 - Y_H}{Y_H} \hat{\mu}_A \times \\ \times (\frac{S_{NH}}{K_{NH} + S_{NH}})(\frac{S_O}{K_{O,A} + S_O}) X_{B,A} \tag{10}$$

2.2. *Parameter Identification*

Very important components of these wastewater treatment process mathematical models are models parameters: reaction rates, saturation coefficients, specific growth rates, *etc.*, which are unknown.

Model structures, experiment planning, and identification condition analysis are applied in classic theory of identification of the model parameters. The mathematical models of wastewater treatment processes are of high complexity; hence, they are amenable only to numerical methods in order to get a solution. Therefore, numerical estimation of parameters could be used.[3]

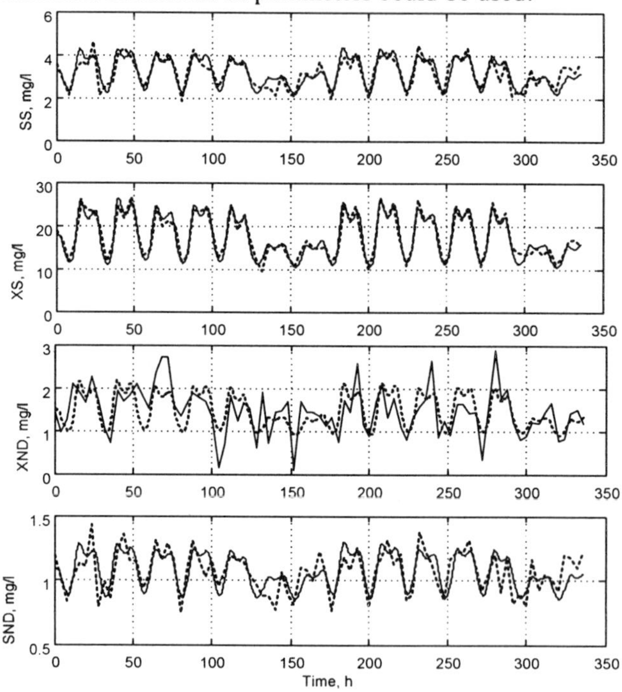

Figure 1. Some output components concentrations. Dashed line- experimental data, continuous line- model outputs.

We used the special optimisation functions of MATLAB for the estimation of wastewater treatment processes mathematical models parameters. Data for the mathematical model identification of the activated sludge process was taken from internet data base.[4] There are presented measured influent and effluent wastewater parameters values of Uppsala wastewater treatment plant of two weeks every fifteen minutes in these data base.

Table 1. Parameter values of the activated sludge process model.

Symbol	Model parameters	Unit	Value
Y_H	Heterotrophic yield	g cell COD formed (g COD oxidized)$^{-1}$	0.67
Y_A	Autotrophic yield	g cell COD formed (g N oxidized)$^{-1}$	0.23
f_P	Fraction of biomass yielding particulate products	dimensionless	0.06
i_{XB}	Mass N/mass COD in biomass	g N (g COD)$^{-1}$ in biomass	0.08
i_{XP}	Mass N/mass COD in products from biomass	g N (g COD)$^{-1}$ in endogenous mass	0.086
$\hat{\mu}_H$	Heterotrophic max. specific growth rate	day^{-1}	0.62
b_H	Heterotrophic decay rate	day^{-1}	1.0
K_S	Half- saturation coefficient (hsc) for heterotrophs	g COD m^{-3}	20
$K_{O,H}$	Oxygen hsc for heterotrophs	g O_2 m^{-3}	0.2
K_{NO}	Nitrate hsc for denitrifying heterotrophs	g NO_3.N m^{-3}	0.47
$\hat{\mu}_A$	Autotrophic max. specific growth rate	day^{-1}	6.0
b_A	Autotrophic decay rate	day^{-1}	0.4
$K_{O,A}$	Oxygen hsc for autotrophs	g O_2 m^{-3}	0.19
K_{NH}	Ammonia hsc for autotrophs	g NO_3.N m^{-3}	0.8
η_g	Correction factor for anoxic growth of heterotrophs	dimensionless	0.8
k_a	Ammonification rate	m^3 (g COD day)$^{-1}$	0.08
k_h	Max. specific hydrolysis rate	g slowly biodeg. COD (g cell COD day)$^{-1}$	3.0
K_X	Hsc for hydrolysis of slowly biodeg. substrate	g slowly biodeg. COD (g cell COD)$^{-1}$	0.03
η_h	Correction factor for anoxic hydrolysis	dimensionless	0.4
$S_{O,sat}$	Saturation DO concentration	g/m^3	10.0

2.3. *Simulator*

The different mathematical models were combined into the complex simulator of the wastewater treatment plant. The software package SIMULINK (MATLAB) was used for realization of the complex mathematical model.

The designed bioreactor with two cameras (volume is 13920 and 14100m^3) was simulated under anaerobic and aerobic conditions (with the recirculation). The complex simulator of the wastewater treatment processes can help to analyze different interferences in wastewater treatment plant input. It is very important, because some processes are lasting weeks or month, so the plant engineers can forecast the processes in the bioreactors, the digesters after long time. The simulator could be used for analyses of control systems.

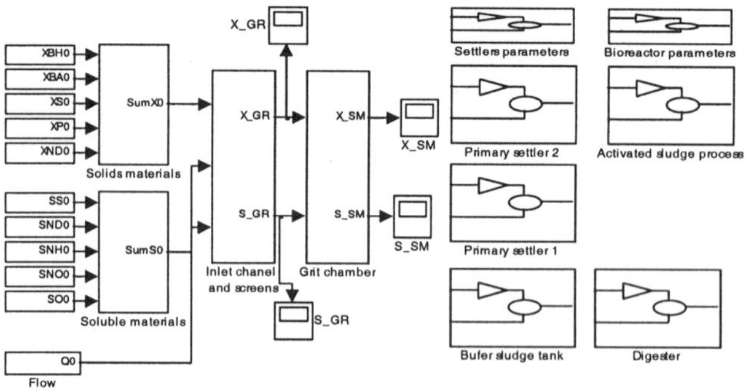

Figure 2. Realization of mathematical model in Matlab/Simulink environment.

3. Process Control

Increasing demands on effluent wastewater quality and increasing loads call for improved control and optimization of WWTP. Applied research in intelligent process control is very important in achieving high overall plant performance. The main objective for control of WWTP is to achieve maximum purification at minimum costs- energy and chemical additions in order to maintain relevant microbial populations within the plant.

A lot of control engineers and scientist are interesting in wastewater treatment process control. Dissolved oxygen (DO) control does not require any in-depth knowledge of the microbial dynamics. Therefore, a traditional PI controller or on/off controller has been widely used[5] and there have been extensive experiences of DO control with feedback controller.[6] Despite the straightforward task of DO dynamics, several difficulties are involved: DO dynamics contain both nonlinear and time varying properties.

Nitrogen and phosphorus are the principal nutrients of concern in treated wastewater discharges Therefore, the control of nitrogen and phosphorus is becoming increasingly important in water quality management and in the design of WWTP.[6,7] There are only a few ways to influence the nitrification/denitrification rates in practice: one is to adjust the DO set point for ammonia removal in the aeration zone; another is to control the dosage of external carbon for nitrate removal. Phosphorus can be removed by controlling the dosage of chemicals for phosphorus precipitation. Yuan *et al.*[8] suggested various control strategies using proportional feedback controller.

Some examples of multivariable control of the wastewater treatment process can be found in Bastin and Dochain.[9] Lindberg[6] suggested multivariable modeling and control strategy of nutrient removal in WWTP using numerical algorithms for subspace state space system identification that can identify multivariable processes. From the multi-input multi-output process model, he developed a linear quadratic (LQ) controller with integration of feedforward and feedback controller. Steffens and Lant[10] evaluated several multivariable model-based control algorithms, such as linear quadratic controller (LQC), dynamic matrix controller (DMC) and nonlinear predictive controller (NPC) for controlling nitrogen removal in WWTP and compared that with a conventional PI controller of the system. They concluded that model-based control algorithms could provide tight control of nitrogen compound removal and offer significant benefits in terms of deferred capital expenditure.

3.1. *GMC for Autotrophic Biomass Specific Growth Rate Control*

The specific growth rate of autotrophic biomass is impact the nitrogen removing process. An accurate control of the specific growth rate of autotrophic biomass is very important. The control is implemented by changing the airflow supplied to the aerobic bioreactor. A model based control approach, known as Generic Model Control (GMC), is studied with respect to its application for the regulation of the specific growth rate of autotrophic biomass. In GMC theory,[11] the non-linear process model is directly embedded into the control law. One of the most attractive features is that this control scheme solves an optimization problem in one-step of calculation only. One of the major drawbacks, however, is the need to have all the states of the process available on-line in order to calculate the control law. Consider the situation in which the dynamics of a process are presented by differential equation:

$$\frac{dy}{dt} = f(y,u,d,t) \tag{11}$$

where y is the process output, u - is the input, d - is the disturbance, and f can be a nonlinear function. Further, consider that we wish for the process output to approach its desired value y_s along a reference trajectory y_r, determined by (16). The process model (15) may be used directly to obtain the controller required to cause the process to follow this trajectory to obtain:

$$\frac{dy_r}{dt} = \frac{dy}{dt} = K_1(y_s - y) + K_2 \int (y_s - y)dt \tag{12}$$

In the case of autotrophic biomass specific growth rate μ_A control, we have changed trajectory of set point $\mu_{A,s}$ to derivative

$$\frac{d\mu_{A,s}}{dt} = K_1(\mu_{A,s} - \mu_A) + K_2 \int_0^t (\mu_{A,s} - \mu_A)dt \tag{13}$$

where K_1 and K_2 are constants, and $\mu_{A,s}$ is the set point of the autotrophic biomass specific growth rate.

The processes in the bioreactor are described in section 2.1. Dissolved oxygen concentration non-linearly depend on airflow (manipulated variable) u(t). Therefore we used the new variable $S_{Oras}(t)$:

$$S_{Oras}(t) = K_L a(u(t))(S_{O\,sat} - S_O) \tag{14}$$

The typical oxygen transfer function is:[9]

$$K_L a(u(t)) = k_1(1 - e^{-k_2 u(t)}) \tag{15}$$

So, estimated airflow is

$$u(t) = -\frac{1}{k_2}\ln\left(1 - \frac{S_{Oras}(t)}{k_1(S_{Osat} - S_O)}\right) \tag{16}$$

where

$$\begin{aligned} S_{Oras}(t) = {} & \frac{k_1(\mu_{A,s} - \mu_A) + k_2 \int_0^t (\mu_{A,s} - \mu_A)dt}{\hat{\mu}_A \dfrac{S_{NH}}{K_{NH} + S_{NH}} \cdot \dfrac{K_{O,A}}{(K_{O,A} + S_O)^2}} - \frac{S_O K_{NH}(K_{O,A} + S_O)}{S_{NH} K_{O,A}(K_{NH} + S_{NH})}\left(\left(-i_{XB}\hat{\mu}_H \frac{S_S}{K_S + S_S} \times \right.\right. \\ & \times \left(\frac{S_O}{K_{O,H} + S_O} - \frac{\eta_g K_{O,H}}{K_{O,H} + S_O}\frac{S_{NO}}{K_{NO} + S_{NO}}\right) + k_a S_{ND}\Big)X_{B,H} - \frac{Q}{V}(S_{NH_{in}} - S_{NH})\Big) - \frac{Q}{V}(S_{O_{in}} - S_O) + \\ & + \mu_{A,s}\left(\frac{4.57 - Y_H}{Y_H} + \frac{(i_{XB} + \frac{1}{Y_A})S_O K_{NH}(K_{O,A} + S_O)}{S_{NH} K_{O,A}(K_{NH} + S_{NH})}\right)X_{B,A} + \hat{\mu}_H \frac{1 - Y_H}{Y_H}\frac{S_S}{K_S + S_S}\frac{S_O}{K_{O,H} + S_O}X_{B,H} \end{aligned} \tag{17}$$

Generic control system and process reaction are presented below.

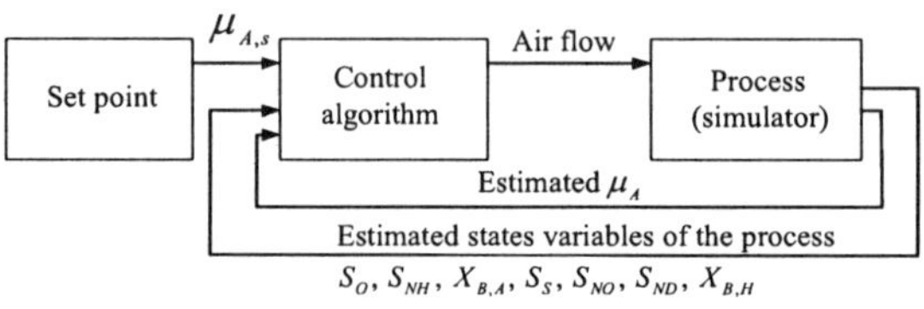

Figure 3. Generic control system.

3.2. *Application of the GMC for Autotrophic Biomass Specific Growth Rate Control*

In aid of the complex WWTP simulator, economy efficient was analysed if autotrophic biomass specific growth rate set point is increased at night and decreased at day time (night and day electrical energy tariffs). There are two tariff rates of electrical energy, which depend on day/night time in Lithuania. The GMC was used for the autotrophic biomass specific growth rate increasing in night time and decreasing in day time during seven days.

The electric energy consumption AE was estimated for Degremont DP230 disk system according to:[12]

$$AE = 24 \cdot \int (0,4032(K_L a)^2 + 7,8408(K_L a))dt \tag{19}$$

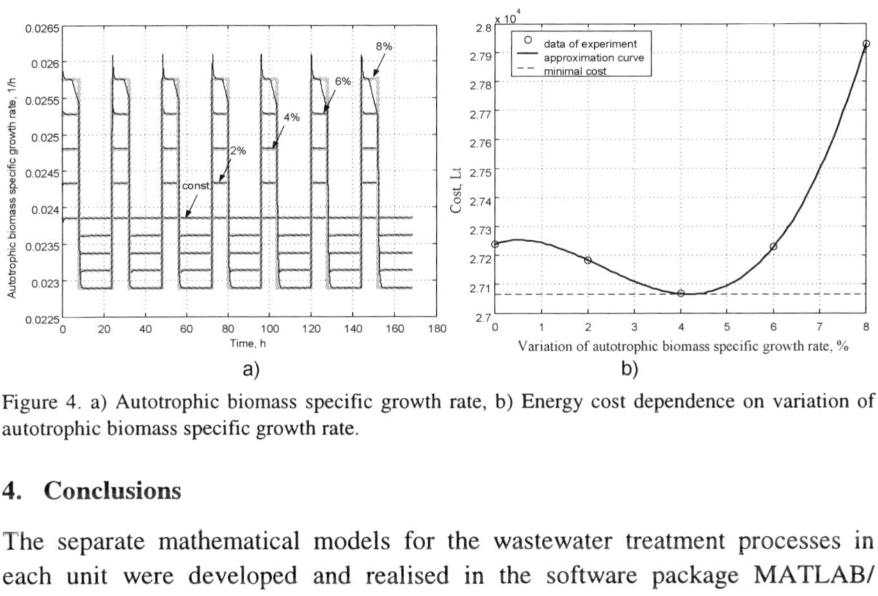

Figure 4. a) Autotrophic biomass specific growth rate, b) Energy cost dependence on variation of autotrophic biomass specific growth rate.

4. Conclusions

The separate mathematical models for the wastewater treatment processes in each unit were developed and realised in the software package MATLAB/ SIMULINK. The different mathematical models were combined into the complex wastewater treatment plant simulator.

GMC was analysed and proposed for the regulation of the specific growth rate of autotrophic biomass. With a help of the simulator, it was calculated that energy cost is reduced by (0.6- 0.8)% if autotrophic biomass specific growth rate set point is increased by 4% at night and decreased by 2% at daytime.

Acknowledgment

The financial support from Lithuanian State Science and Studies Foundation is gratefully acknowledged.

References

1. A. Rehnström, *Automatic Control of an Activated Sludge Process in a Wastewater Treatment Plant, a Benchmark Study*, MSc thesis (2000).
2. J. Repsyte and R. Simutis, Mathematical modelling principles of wastewater treatment processes. *Automation and Control Technologies-2002*, Technologija, Kaunas, 56-61 (2002).
3. B. A. Ogunnaike, W. H. Ray, *Process Dynamics, Modeling, and Control*, New York, Oxford (1994).
4. COST 682/624 Action website (http://www.ensic.u-nancy.fr/COSTWWTP) - The European Co-operation in the field of Scientific and Technical Research.
5. M. J. Flanagan, B.D. Bracken and F. J. Roeler, Automatic dissolved oxygen control, *J. Env. Eng.* **103**, 707 (1977).
6. C. F. Lindberg, Multivariable modeling and control of an activated sludge process, *Wat. Sci. Tech.* **37**, 149 (1998).
7. C. K. Yoo, J. H. Cho, H. J. Kwak, S. K. Choi, H. D. Chun and I. Lee, Closed-loop identification and control for dissolved oxygen concentration in the full-scale coke wastewater treatment plant, *Wat. Sci. Tech.*, in press (2001).
8. Z. Yuan, H. Bogaert, P. Vanrolleghem, C. Thoeye, G. Vansteenkiste and W. Verstraete, Control of external carbon addition to predenitrifying systems, *J. Envir. Engng.* **123**, 1080 (1997).
9. G. Bastin and D. Dochain, *On-Line Estimation and Adaptive Control of Bioreactor*, Elsevier, New York (1990).
10. M. A. Steffens and P. A. Lant, Multivariable control of nutrient removing activated sludge systems, *Wat. Res.* **33**, 2864 (1999).
11. P. L. Lee and G. R. Sullivan, General model control – theory and applications, *IFAC Symposium on Model Based Process Control*, Atlanta GA, USA (1988).
12. http://www.degremont.com